Food, Economics, and Health

Food, Economics, and Health

Alok Bhargava

OXFORD

UNIVERSITY PRESS

Great Clarendon Street, Oxford OX2 6DP

Oxford University Press is a department of the University of Oxford.
It furthers the University's objective of excellence in research, scholarship,
and education by publishing worldwide in

Oxford New York

Auckland Cape Town Dar es Salaam Hong Kong Karachi
Kuala Lumpur Madrid Melbourne Mexico City Nairobi
New Delhi Shanghai Taipei Toronto

With offices in

Argentina Austria Brazil Chile Czech Republic France Greece
Guatemala Hungary Italy Japan Poland Portugal Singapore
South Korea Switzerland Thailand Turkey Ukraine Vietnam

Oxford is a registered trade mark of Oxford University Press
in the UK and in certain other countries

Published in the United States
by Oxford University Press Inc., New York

© Alok Bhargava 2008

The moral rights of the author have been asserted
Database right Oxford University Press (maker)

First published 2008

British Library Cataloguing in Publication Data

Data available

Library of Congress Cataloging in Publication Data

Data available

Typeset by SPI Publisher Services, Pondicherry, India
Printed in Great Britain
on acid-free paper by
Biddles Ltd., King's Lynn, Norfolk

ISBN 978–0–19–926914–3

1 3 5 7 9 10 8 6 4 2

Contents

Preface

This monograph is based on a series of lectures developed for a course on food, economics and health that I have taught at various universities. Many economists and nutritionists interested in food policy issues have encouraged me to turn the material into a monograph because of the cross-disciplinary themes involved. One of the problems in social science approaches to food policies is that students and researchers are often unfamiliar with the literature in nutritional sciences. Similarly, nutritionists and epidemiologists are often not familiar with economics terminology, such as "endogeneity" of explanatory variables, that is relevant, for example, when one tries to explain individuals' nutrient intakes by anthropometric indicators such as heights and weights. Further, the effects of under-nutrition in developing countries on outcomes such as children's growth and cognitive development are recognized to be of utmost policy importance. Due to the obesity epidemic in developed countries, however, researchers also need to be familiar with the effects of diet and lifestyle on weight gain. Moreover, nutrition education programs in developed countries are partly designed by psychologists, so that specification of empirical models for food consumption and interpretation of the results inevitably require multi-disciplinary approaches. While it may seem a complex task to grasp the approaches in economics, nutrition and psychology, basic familiarity with these literatures is helpful for analyses of food consumption and population health data. This monograph attempts to bridge the gaps between different disciplines interested in food policy formulation.

The material covered in the six main chapters emphasizes the use of longitudinal ("panel") data from developing and developed countries for food policy analyses because the relationships between food intakes and health outcomes are complex. This complicates the presentation of methodological issues, and readers not familiar with statistical aspects can skip the details and focus on substantive findings. Moreover, undergraduate and graduate students in economics, agricultural economics, nutrition, demography, public health and multi-disciplinary programs can focus on aspects that are of interest for their disciplines. Readers interested in further details can refer to my published articles reproduced in Bhargava, 2006a. In addition, Chapter 1 in this monograph provides an overview of the contents, and the Glossary summarizes the terminologies used in the several disciplines. Moreover, certain unpublished results on issues of child development and gender bias in developing countries are included in Chapters 4 and 5, respectively. I hope that readers will find the presentation to be a useful introduction to scientific and methodological issues in food and health policy analyses, especially since diets and lifestyles affect the well-being of everyone.

Last, I would like to thank the publishers of the *Journal of the Royal Statistical Society, American Journal of Human Biology, American Journal of Physical Anthropology, Journal of Educational Psychology, Preventive Medicine, Journal of Econometrics* and *Indian Economic Review* for their kind permission to reproduce certain tables from my previously published articles. Thanks are also due to Ms Jennifer Wilkinson of Oxford University Press for her efforts at improving the exposition.

Glossary

24-hour recall	Recollection of all foods consumed in the previous 24-hour period
7-day food record	Record of all foods consumed every day for a 7-day period
anemia	Hemoglobin concentration of less than 12 grams per liter of blood for women (<13 for men)
bioavailable iron	Iron that can be absorbed by the body, e.g. iron from meat, fish and poultry ("heme iron") is highly bioavailable
BMI	body mass index; weight in kilograms divided by height in meters squared
BMR	basal metabolic rate; minimum amount of energy needed for sustaining life
dynamic models	Models containing previous realizations of the dependent variable
elasticity	Percentage change in a dependent variable resulting from a 1% change in an explanatory variable
endogenous variable	A variable that may be correlated with errors affecting the dependent variable
exogenous variable	A variable that is assumed to be uncorrelated with errors affecting the dependent variable
fixed effects models	Models containing an indicator (or dummy) variable for each individual in the longitudinal model

food expenditure surveys	Surveys recording expenditures incurred by households on various food items (on a quarterly or yearly basis)
food frequency questionnaires	Questionnaires concerning the frequencies with which various food groups were consumed in the previous 30 to 90 days
GDP	Gross Domestic Product (per capita)
habit persistence	Dependence of current consumption patterns on past habits
heteroscedasticity	Errors for each individual unit in the model having different variances
income elasticity	Percentage change in a dependent variable resulting from a 1% change in income
instrumental variables	Variables used to predict endogenous explanatory variables included in the model
ICRISAT	International Crops Research Institute for Semi-Arid Tropics
kilocalorie (kcal)	Unit for measuring dietary energy: heat required for raising the temperature of 1 liter of water by 1 degree centigrade
kilojoule (kJ)	0.239 kcal
likelihood ratio tests	Statistical tests for null hypotheses based on maximum likelihood estimates
LSMS	Living Standards Measurement Surveys
macronutrients	Nutrients such as protein, carbohydrates and fats from which most dietary energy is derived
maximum likelihood estimates	Parameters estimated by maximizing the likelihood that the sample came from the postulated distribution
micronutrients	Nutrients such as minerals and vitamins needed in small quantities for maintaining health
NFHS	National Family Health Survey

obese	BMI greater than 30
overweight	BMI greater than 25
random effects models	Models containing randomly distributed variables for each individual unit in the longitudinal model
RDA	recommended dietary allowances
static models	Models not containing previous realizations of the dependent variable as an explanatory variable
stunting	Low height-for-age
wasting	Low weight-for-age (or weight-for-height)
WHTFSMP	Women's Health Trial: Feasibility Study in Minority Populations
z-scores	Deviation of height (weight) from median height (weight) for the age group in the reference population, divided by the standard deviation of height (weight) in the reference population

1

Introduction

1.1 General background

The recent increases in communications, travel and trade among countries, partly due to globalization, have brought about major changes in food consumption patterns of populations. This is especially true of developed countries, where availability of different types of cuisines has led to a type of "convergence in tastes" of populations. However, diets may not be optimal from a health standpoint, partly because of sugar and fat content and also due to increasingly sedentary lifestyles. Moreover, even in middle-income and developing countries, the affluent are consuming a variety of cuisines that are contributing to the obesity epidemic and increasing the prevalence of chronic diseases such as hypertension, diabetes and cardiovascular disorders. At the same time, problems of hunger and poor diet quality are widely prevalent in developing countries. Inadequate food intakes by children hinder physical growth and can impair their mental development; poor quality of diet in terms of low intakes of micronutrients such as calcium, iron and vitamins can adversely affect children's immune systems, thereby hampering school attendance and learning. Children's cognitive development is critical for increasing the future supply of skilled labor and for economic growth (Bhargava, 2001a; Bhargava, Jamison *et al.*, 2001). From a food policy standpoint, policy-makers in developing countries are now faced with the complex tasks of combating diseases associated with under-nutrition as well as those due to over-consumption of food. It is therefore essential for researchers in the social and biomedical sciences to take a broad view of the consequences of sub-optimal diets for population health and well-being.

Historically, nutritionists working in developing countries tackled diseases due to hunger and poverty and were mainly concerned with deficiencies in dietary energy, protein and micronutrients (see e.g. Scrimshaw *et al.*,

1959; Waterlow, 1974). Nutritional epidemiologists in the latter half of the twentieth century turned their attention to the adverse effects of excessive intake of dietary fat and cholesterol on chronic conditions such as cardiovascular disease and cancers (Keys, 1980, 1984). As information from various studies continues to accumulate, research in food policy issues can incorporate the relevant knowledge to develop effective policies for disease prevention via dietary modifications. One of the earliest applications in food policy analysis was the influential work by Hutchinson (1969) that underscored the importance of food shortages ("energy deficiencies"), noting that "protein when divorced from carbohydrate in the diet is of no use in repairing the wear and tear of body protein". Thus, policy-makers should address energy deficiencies before combating protein deficiencies. Knowledge from the nutritional sciences is useful for formulating food policies that can potentially affect the lives of millions of individuals in making healthful food choices. Moreover, economic analyses have a central role in policy formulation because food prices and household incomes affect food consumption decisions, especially in developing countries. Changes in diet and lifestyles are also important and analyses should incorporate behavioral (i.e. psychological) factors for enhancing the efficacy of educational programs encouraging healthful eating.

The material presented in this monograph can be divided into four broad themes. First, evidence from developing countries on the effects of household incomes and other variables on food consumption patterns is summarized in Chapter 2. Second, the effects of food intakes on child health outcomes such as height, weight and morbidity in developing countries are covered in Chapter 3; nutritional and other factors affecting children's cognitive development are addressed in Chapter 4. Because high fertility (birth) rates affect surviving children's health status, issues of child survival typically covered in demographic research are addressed in Chapter 5. Third, links between adults' nutritional and health status and their economic (labor) productivity in developing countries are addressed in Chapter 6. Fourth, owing to excess food consumption, especially in developed countries, an obesity epidemic is increasing the prevalence of chronic diseases. Chapter 7 summarizes diverse approaches to obesity research in the fields of nutrition, psychology and economics. It should be emphasized that a major objective of this book is to provide an overview of quantitative analyses of issues surrounding food consumption and population health. Because researchers in disciplines such as economics, nutrition, psychology, demography, anthropology and public health emphasize somewhat different aspects, this presentation should be useful for assessing the relative importance of the factors affecting population health. Readers less familiar with statistical methods can skip technical discussions in many places and focus on substantive issues relating to food consumption and population health; the glossary summarizes some of the terminology used in different disciplines.

An important aspect of analyzing the links between food intakes and health outcomes is that food consumption decisions are heavily influenced by customs, especially in traditional societies. Because such customs have evolved gradually over time, individuals are likely to change their food consumption patterns slowly. By contrast, rapid industrialization has led to dramatic changes in food intakes. From a health standpoint, however, it is not obvious that rapid changes in diets afforded by economic development are preferable to changes occurring gradually over time. Moreover, traditional diets do not necessarily imply poor understanding of nutritional issues even among populations with low literacy rates; diets have evolved by observing the health benefits of intakes of essential foods such as grains, milk and meat. Moreover, the practice of vegetarianism in India was partly based on religious philosophy and also on the expected workload of individual groups ("castes"). For example, the upper castes typically did not perform strenuous work and did not appear to require nutritious foods such as meat that were deemed essential for "warrior" castes. From a modern perspective, while vegetarian diets may supply inadequate quantities of vital nutrients such as iron that can be absorbed by the body, it is also the case that excessive fat and cholesterol intakes in industrialized societies are deleterious for health. Both situations, however, can be remedied by appropriate nutritional and/or educational interventions. For example, iron supplementation and increased consumption of dairy products can reduce anemia prevalence among vegetarians, while increasing consumption of whole grain products and fruits and vegetables can lower the fat and cholesterol content of Western diets.

In developing a framework for analyzing the effects of food intakes on health outcomes, two sets of issues need to be addressed. First, problems of energy and micronutrient deficiencies should be urgently tackled in developing countries. Combating these deficiencies requires elaborate policies, especially in areas of food shortages or where the price system may inhibit consumption of foods high in nutrients such as protein, iron and vitamins A and C. Moreover, even in developed countries such as the US, children from poor backgrounds can face hunger spells and poor households benefit from programs such as the Supplemental Food Program for Women, Infants, and Children, which supplies milk, cheese and fruit juice to pregnant women and to children less than 5 years of age. The second point is that because of unhealthy eating patterns, high levels of economic development do not necessarily entail improved dietary intakes. In addition, sedentary lifestyles associated with affluence reduce physical activity, thereby throwing individuals into a positive energy balance, i.e. promoting weight gain. Because these problems are also prevalent in middle-income countries and are on the rise in developing countries, it is fruitful to analyse the effects of sub-optimal diets on individual health using a comprehensive modelling framework.

Further, there is a need to integrate the literatures in the nutritional and social sciences for understanding the effects of food intakes on health, though the task is a complex one. For example, there are debates in the nutritional sciences regarding the effects of dietary fat on obesity (e.g. Willett, 1998a; Bray and Popkin, 2000). If one adds to such controversies the different approaches to food consumption in disciplines such as anthropology, psychology and economics, then developing a common analytical framework seems an elusive task. However, certain processes are well understood in the biomedical sciences and it is important to incorporate that knowledge in empirical analyses of food intakes and health outcomes. For example, studies such as Spurr (1993) have demonstrated that oxygen uptake, reflecting physical work capacity, is reduced in under-nourished individuals with low body mass index (BMI). Similarly, anemic individuals with low hemoglobin concentration were found to take longer to perform agricultural tasks in Indonesia (Basta *et al.*, 1979), which is likely to be due to reduced oxygen transportation. Not surprisingly, therefore, an analysis of time allocation patterns among Rwandese adults showed that individuals with a low BMI participated in less strenuous activities and spent greater amount of time resting and sleeping (Chapter 6). While knowledge from scientific experiments is useful for specification of empirical models for health outcomes, the pathways are often complex. For example, the link between low micronutrient intakes and children's scores on cognitive tests in developing countries is complicated by the fact that the school environment plays an important role in child development. Such issues are addressed in Chapter 4 using data from developing countries. A multidisciplinary approach incorporating knowledge from the biomedical and social sciences is useful for modeling the pathways through which nutritional and health status affect critical outcomes such as children's cognitive development and the labor productivity of adults.

Economists have been interested in issues of food, nutrition and health policies for a variety of reasons. In developing countries, food prices play a fundamental role in ensuring adequate intakes; smooth functioning of agricultural markets is essential for the well-being of a large number of individuals. From a scientific standpoint, it is also critical to understand the effects of specific nutrient intakes for maintaining individual health and productivity. For example, Stigler (1945) estimated "minimum cost diets" by the linear programming method, though noting the importance of interactions between nutrients present in the meal for nutrient absorption. While nutrient interactions are complex processes studied in laboratory settings, insights from experiments can be valuable for formulation of food policies. A good example is the widely prevalent iron deficiencies in less developed countries due to low intakes of absorbable iron (UNICEF/ WHO, 1999). Iron from meat, fish and poultry ("heme iron") is more easily absorbed by the human body than non-heme iron derived from staple foods such as wheat and rice; the presence of meat and ascorbic acid in the meal increases the absorption of non-heme iron (Monsen *et al.*, 1978). Thus, an important issue for food policies is whether absorption of iron from staple foods can be enhanced by increasing the iron content of staple foods and/or by higher intakes of enhancers of iron absorption such as meat and ascorbic acid (Bhargava, Bouis *et al.*, 2001). Without an understanding of the nutritional issues, it would be difficult to formulate such a policy discussion. A multi-disciplinary approach to food policies is likely to suggest cost-effective ways of reducing deficiencies of iron and other vital nutrients in developing countries.

An important example linking knowledge of the biochemistry of food to economic policy is the work by Leibenstein (1957) arguing that higher wages for workers in developing countries would improve their nutritional status and hence productivity. This "wage efficiency hypothesis" has influenced work by economists such as Mirrlees (1975), Stiglitz (1976), Bliss and Stern (1978) and Dasgupta (1993). In practice, the relationship between energy intakes and wages is affected by factors such as the need of workers to spend part of their wages to support family members (Majumdar, 1959), demand conditions in the regional labor markets, micronutrient deficiencies such as iron deficiencies that can depress physical work capacity, and the time necessary for improving individuals' health status. Furthermore, nutritionists such as Beaton (1984) have argued that lower food intakes can lead to reductions in energy expended on tasks, i.e. induce behavioral and economic changes. Thus, it is useful to analyze the effects of food shortages by investigating the effects of nutritional status on individuals' time allocation patterns. Chapter 6 summarizes some of the literature on relationships between health and wages in developing countries. By combining the literatures in nutrition, economics and health, one can enhance the substantive content of empirical models and arrive at policy conclusions that can benefit large numbers of individuals subsisting on agriculture.

Another example for the need for adopting a multi-disciplinary approach to formulating food policies is the obesity epidemic, affecting population health especially in developed countries. Food budgets are relatively small proportions of total household budget in developed countries because of low food prices and higher incomes. In addition, low wages in the service sector depress food prices, especially in fast-food restaurants. Such restaurants often increase the fat content of prepared food to appeal to customers' tastes and may promote over-eating by offering larger portion sizes. While nutritionists are interested in issues such as the effects of specific nutrients like saturated fat on body weight, it is likely that economic factors such as low food prices and profit margins are contributing to the obesity epidemic. Moreover, it is difficult for economists to devise interventions for promoting healthy eating without knowledge of nutritional issues underlying the relationships between diet and obesity. By the same token, nutritionists and epidemiologists spend considerable effort making links between fats, carbohydrate and protein intakes and adiposity (Bhargava and Guthrie, 2002), though in short time frames. Because such analyzes cannot address the underlying economic factors that contribute gradually to the obesity epidemic, a multi-disciplinary approach is essential for understanding the effects of food consumption and other variables on population health. The motivation for organizing the material in the ensuing chapters is outlined in sections 1.2–1.5.

1.2 Economic factors and energy and micronutrient deficiencies in developing countries

Historically, shortages of staple foods such as wheat and rice were prevalent in developing countries, leading to hunger and starvation (Dreze and Sen, 1990; Fogel, 1994). Thus, selling staple foods at subsidized prices through special shops has been a popular strategy of governments in developing countries for reducing under-nutrition. There is a literature in agricultural economics on the formulation of food policies for developing countries (e.g. Timmer *et al.*, 1983; Pinstrup-Andersen, 1988), in part because subsidies for staple foods can constitute a large proportion of national budgets. In cases of food shortages, food imports and the food distribution system play a critical role in alleviating hunger; wars and civil strife can hamper food distribution. Although recent increases in food production have reduced food shortages, deficiencies of protein and micronutrients such as vitamins and minerals are widely prevalent in developing countries (IFPRI, 1990). For example, shortfalls in vitamin A and C intakes in Indonesia were found to contribute to nutritional blindness (Sommer, 1990). Because the productivity of the future labor force depends on children's learning (Bhargava, 2001a), and adult productivity is adversely affected by iron deficiencies (Basta *et al.*, 1979), it is perhaps somewhat narrow from a food policy standpoint to focus only on the adequacy of energy intakes. Instead, policy-makers should adopt the broader goal of improving "quality of diet" in developing countries.

The effects of household incomes on food intake are important for food policy analyzes. However, most researchers in the past analyzed data from household expenditure surveys on aggregate commodity groups such as food, clothing and housing. Because of the high degree of aggregation, one can over-estimate magnitudes of the effects of income on energy and nutrient intakes using these types of data. The effects of household incomes on diet quality are particularly informative when individuals' food intakes are directly measured over time, i.e. when longitudinal data on food intakes and socioeconomic variables are available for analysis. For example, a longitudinal study in three states in south India conducted by the International Crops Research Institute for Semi-Arid Tropics (ICRISAT; Ryan *et al.*, 1984), and a study in the Mindanao region of the Philippines (Bouis and Haddad, 1992) have provided valuable dietary, health and socioeconomic information. Because of the logistics of monitoring households over time, such studies are often confined to specific geographical regions. Thus, households face similar food prices, and issues such as the effects of household incomes on quality of diets in terms of micronutrient content can be analyzed in a more systematic manner than was previously feasible. When longitudinal data on food intakes are available, the likely effects of food subsidies on health outcomes, such as anthropometric indicators, discussed in the earlier food policy literature can also be analyzed using more complex formulations.

Chapter 2 covers issues surrounding the effects of household incomes on food consumption in developing countries. The discussion recognizes the importance of cultural factors affecting food consumption and postulates "dynamic" demand models that are consistent with "habit persistence" in diets (Gorman, 1968). Moreover, the nutrition literature is integrated into the presentation and it is underscored that social scientists need to understand certain basic aspects of the nutrition literature. For example, most nutrients are present in most foods though in different proportions. Thus, while increases in food intake generally increases the intake of nutrients such as protein, iron, calcium and vitamins A and C, choices between different foods entail large changes in the distribution of nutrient intakes. Individuals desiring greater quantities of vitamins A and C, for example, are likely to consume more fruits and vegetables, though such foods also contain small quantities of protein and iron. This integrated approach to food consumption is in the spirit of Stigler (1945), though it differs from contributions in the economics literature in that simplifying assumptions which may be inconsistent with biomedical knowledge are not invoked. Policy-makers are more likely to incorporate research findings into food policy formulation if analyses are based on scientifically acceptable assumptions.

Another aspect of the material covered in Chapter 2 is that it is recognized that individuals' energy and nutrient requirements influence their food intakes. Thus, it is important to account for individuals' heights and weights in models for energy and nutrient intakes. The discussion can be viewed as extending the previous literature in food economics, since section 2.3 first presents models for demand for food at the household level. Moreover, previous estimates of the effects of household incomes on energy and nutrient intakes ("income elasticities") from food expenditure data at the household level are summarized. Subsequently, section 2.4 presents models for energy and nutrient intakes estimated by using data on individuals' intakes; it is argued that one would expect income elasticities based on food intakes to be lower than those estimated by using expenditure data at the household level. Moreover, small magnitudes of income elasticities of energy (~ 0.1) and nutrients estimated from individual-level data are consistent with biomedical evidence on individuals' requirements; it is perhaps unrealistic to expect income elasticities of energy to be very high (~ 0.8) unless there are severe food shortages and starvation. The empirical estimates of income elasticities for energy and nutrients from India, the Philippines, Bangladesh and Kenya are discussed in section 2.5. The econometric models employed are similar and hence the estimates provide interesting contrasts between these countries. For example, because income levels were low in Kenya, one would expect the income elasticity of energy to be higher in Kenya (0.29) than (say) in the Philippines, where it was 0.08. The analytical framework and empirical results in Chapter 2 should be useful for the formulation of food policies that seek to ensure adequate energy and nutrient intakes in developing countries.

1.3 Food policies and children's physical and mental development in developing countries

A major goal for food policies is to facilitate children's physical and mental development in less developed countries. The formulation of food policies that improve the protein and micronutrient content of diets is important for enhancing children's anthropometric indicators. Moreover, the effects of micronutrient deficiencies such as those of iron on children's cognitive development have been emphasized by psychologists and nutritionists (e.g. Pollitt *et al.*, 1993; Scrimshaw, 1996). However, the literature in the biomedical sciences has mainly estimated simple correlations or associations between dietary intakes and cognitive test scores. At low levels of income, children's physical development and learning are intertwined; children's cognitive development depends on their nutritional and health status, household resources, and the educational infrastructure (Bhargava, 1998; Bhargava, Jukes *et al.*, 2005). It is therefore important for social scientists and policy-makers to investigate the proximate determinants of children's cognitive development, which is critical for the future supply of skilled labor and hence for economic development.

Chapter 3 begins by discussing nutritional and socioeconomic factors affecting children's physical development and morbidity in developing countries. Methodological issues such as the implications of the "health production functions" terminology used in the economics literature are spelled out; it is noted that "health functions" is a more general concept because of differences between production of commodities and the manner in which children achieve better health status. Econometric models for children's heights, weights and morbidity are developed in section 3.2 and graphically illustrated in Fig. 3.1. These models reflect knowledge in nutritional sciences, such as the beneficial effects of higher protein intakes on children's heights and weights. Moreover, the relationships between height and weight emphasized in the anthropometric assessment literature are incorporated into the models. It is also noted that these models for children's anthropometric indicators and morbidity are consistent with "stock/flow"-type formulations used in the economics literature.

Section 3.3 presents the results from models for heights and weights of children estimated using data from the Philippines and Kenya. The models show beneficial effects of diet quality on children's anthropometric indicators. The models for child morbidity are presented in section 3.4 and the findings are likely to be useful for the formulation of food policies. For example, higher intakes of vitamin A were associated with lower child morbidity in the Philippines and Kenya. Moreover, for Kenyan children, data were available on the scores obtained by parents on cognitive tests. The results showed that while parental years of education were not significant predictors of lower child morbidity, parental test scores were significant predictors. Thus, education programs improving, in particular, maternal knowledge regarding children's diets and health are likely to lower morbidity in countries with low female education. Chapter 3 also discusses certain studies investigating the determinants of child morbidity in Bangladesh, Pakistan, the Philippines and Peru.

Chapter 4 is concerned with the effects of children's nutritional and health status on cognitive development and begins with analysis of the effects of maternal nutritional status on infant health. Some of the biomedical literature is briefly summarized and the results from modeling the effects of maternal nutritional status on Kenyan infants' length, head circumference and weight at birth are presented. Moreover, maternal health status plays an important role in infant growth between 1 and 6 months, as is shown by dynamic econometric models for length, head circumference and weight. For example, maternal hemoglobin concentration and BMI were important predictors of the dynamics of Kenyan infants' weights in the first six months of life.

Further, Chapter 4 addresses children's physical growth and cognitive development in developing countries. An important randomized controlled trial in Guatemala providing a nutritional supplement is described and the results are critically assessed. Section 4.4 discusses the implications of the theoretical contributions of Vygotsky (1987) to the developmental psychology literature for empirical models for child development. The analytical framework is useful for understanding potentially differential effects of children's nutritional status on their scores on verbal and analytical tests. Evidence is presented on Kenyan school children and the results were broadly in accordance with theoretical considerations. For example, while children's BMI and head circumference were important predictors of the scores on digit span, Raven's matrices and arithmetic, the anthropometric indicators were less important for explaining scores on verbal tasks. Furthermore, analyses of data from a randomized controlled trial in Tanzania treating school children against hookworm and schistosomiasis infections are presented in section 4.5. Methodological issues such as the importance of modeling the pathways underlying children's cognitive development are discussed. These issues are important due to the current popularity of randomized trials in social sciences. For example, it is argued that data from the treatment group are important for assessing the benefits of removing hookworm and schistosomiasis for children's iron status, which is reflected in hemoglobin and ferritin concentrations. By contrast, children's scores on cognitive and educational achievement tests are likely to evolve gradually and are also influenced by the school infrastructure. Thus, modeling test scores using data from the control group can provide useful insights. For example, better-qualified teachers and greater numbers of work assignments in Tanzania were found to enhance children's scores on educational achievement tests.

Chapter 5 covers demographic, health and mortality issues in developing countries. While nutritional supplementation programs for pregnant women are likely to show beneficial effects on child growth, most studies cover only a small number of women. Large-scale interventions are expensive and often formulated using piecemeal approaches. More importantly, many pregnant women receive no ante-natal care and children are often not vaccinated due to poor access to health care. In the absence of family planning methods, large numbers of children born to women within short time intervals are detrimental for maternal and child health. As documented in Chapter 4, children with poor nutritional and health status are unlikely to reach their full potential. Thus, children's health status and cognitive development also depend on the health care and family planning services available. It is important for formulating cost-effective policies to recognize the synergisms between food and health care policies. Knowledge from disciplines such as nutrition, psychology, demography and economics can be combined for enhancing child health in developing countries.

Chapter 5 begins by discussing the possible discrimination against girls in south Asia emphasized in the demography and economics literatures (e.g. Chen *et al.*, 1981; Sen and Sengupta, 1983). Some methodological difficulties in making comparisons using children's heights and weights are illustrated, using the data from Pakistan and Vietnam. It is argued that child mortality is a more suitable indicator for assessing gender discrimination. Proximate determinants of child mortality are briefly discussed in section 5.2, using national averages for countries. Section 5.3 presents an analysis of infant mortality in Uttar Pradesh using household data from the National Family Health Survey-1 to investigate the possible reasons for higher mortality among girls. Methodological issues such as the dynamic relationships between fertility (number of children) and child mortality, and certain econometric aspects, are discussed. The empirical results showed that the chances of female mortality were higher for "unwanted" girls born at high birth orders (e.g. the fifth child). In contrast, the mortality chances of girls born at low birth orders were *lower* than for boys. The results also showed the importance of health care infrastructure; for example, women vaccinated against tetanus experienced significantly lower infant mortality. These results underscore the importance of ante- and post-natal care for women and infants in enhancing children's survival chances and for improving the health status of the surviving children, which is critical for learning.

Lastly, Section 5.4 analyses the data from another demographic survey from Uttar Pradesh (PERFORM), which compiled extensive information on the health care infrastructure. For example, there are numerous private and public health care providers in countries such as India and some may have very limited qualifications. Because the PERFORM survey compiled detailed information, the effects of health care utilization on fertility can be analyzed. Moreover, from a conceptual standpoint, the frameworks in the economics of supply and demand for children (Easterlin and Crimmins, 1985) and that of "endogenous placement" of health care facilities are reappraised. The empirical results for models for contraceptive use, such as female sterilization, the intrauterine device, birth control pills and condoms, indicate the importance of the health care infrastructure in public and private facilities. Overall, it is emphasized in Chapter 5 that tackling demographic issues such as high fertility and child mortality rates in developing countries is important for improving the well-being of women and children and for creating skilled labor, which is critical for economic growth.

1.4 Nutritional status and labor productivity in developing countries

The links between the nutritional and health status of adults and labor productivity are of fundamental importance to policy-makers, especially in developing countries. Moreover, unlike for children, the effects of improving adult health status are visible in a short time. For example, treatment of adults with full-blown AIDS can lead to dramatic improvements in their health and a return to employment. This is especially the case in developed countries, where HIV-positive individuals can remain productive for years. By contrast, labor productivity may be hindered by co-morbidities and nutrient deficiencies, especially in developing countries. While adult health in developing countries is an important determinant of performance in physically demanding tasks, from a food policy viewpoint it is important to estimate the costs and benefits of improving diets.

The literatures in the biomedical sciences and economics on the effects of health status on labor productivity have emphasized somewhat different aspects. In the biomedical sciences, individuals' physical work capacity is assessed using tests of endurance that provide quantitative measures of performance (e.g. Spurr, 1983). Biomedical scientists are interested in the effects of variables such as adults' height, muscle mass and hemoglobin concentration on the performance of physical tasks. By contrast, the economics literature on the "wage efficiency hypothesis" has emphasized the effects of higher wages on labor productivity via the effects on improved nutritional intakes. At a conceptual level, biomedical studies are simpler to interpret since they are based on well-understood principles of human physiology. In contrast, social science phenomena are more complicated since wages depend on labor market conditions and the effects of higher wages on health and productivity are likely to operate with delays. Chapter 6 outlines the biomedical and economics approaches to analyzing relationships between adult health and productivity and points out difficulties in interpreting empirical results because many adults in developing countries do not receive monetary wages.

The background literature in the biomedical sciences on the effects of nutritional and health status is briefly summarized in section 6.1. A randomized controlled trial of iron supplementation for rubber plantation workers in Indonesia (Basta *et al.*, 1979) is discussed in detail. Moreover, interpretation of the findings from randomized trials is discussed. It is argued that when the scientific phenomena are well understood, it may not be necessary to conduct randomized trials to accumulate further evidence such as that on the benefits of iron supplementation on labor productivity. Instead, policy-makers can utilize the results from previous randomized trials to devise food policies for improving productivity. While the benefits of iron supplementation in different settings would depend on "habitual" dietary intakes and iron loss due to intestinal parasite infections, resources available for research can be used for assessing nutrient deficiencies in the populations and for designing cost-effective policies.

Section 6.2 discusses empirical models used in the economics literature for analyzing the relations between health and wages in developing countries. Moreover, Figure 6.1 summarizes the relationships between wages, food intakes, anthropometric measures, and physical work capacity together with underlying variables that can modify these relationships. The long-run effects of better nutritional and health status are emphasized in the discussion. Section 6.3 summarizes the results from a study of time allocation patterns in Rwanda which showed beneficial effects of adults' BMI and household intakes on time allocated to productive activities (Bhargava, 1997). Moreover, under-nourished adults were found to spend greater proportions of time resting and sleeping. While the effects of health on wages were also investigated in this study, the main advantage of analyzing time allocation patterns is that the relationships are not confounded by market forces affecting the demand and supply of labor. Overall, the results in Chapter 6 show the benefits of improving adults' nutritional and health status on labor productivity. The extent to which different foods that are good sources of energy, protein and micronutrients should be subsidized would depend on food prices and the budget for subsidies. It is noted in Chapter 2 that energy deficiencies are less common outside sub-Saharan Africa and policy-makers need to estimate the prevalence of protein and micronutrient deficiencies for designing cost-effective food policies.

1.5 Diet and obesity in developed countries

The final chapter in this book is concerned with issues of behavior, diet and obesity in developed countries such as the US. There is a stark contrast between food choices made by the poor in developing countries struggling to meet their energy and nutrient requirements, and excess food consumption in developed countries. Moreover, even in developed countries, low-income households may face food shortages, while the affluent in developing countries are beginning to suffer from chronic diseases such as diabetes and coronary heart disease due to excess weight. Thus, policy-makers need to address simultaneously issues of under-and over-nutrition, and the design of food policies is complicated by the dependence of food consumption patterns on behavioral (psychological) and economic factors. For example, subsidies for agriculture in developed countries lower food prices and can promote excess food consumption, while jeopardizing the livelihoods of small farmers in developing countries that cannot compete with large-scale mechanized agriculture.

Section 7.1 begins by summarizing the nutritional and epidemiological approaches to obesity. In these disciplines, hypotheses driving the investigations are well defined and randomized controlled trials are often conducted to test theories. For example, an important hypothesis is that intakes of dietary fat can promote obesity because fat is poorly oxidized, especially by sedentary adults. While this hypothesis has been supported in laboratory experiments, in practice, fat is energy-dense and contains 9 kilocalories (kcals) per gram. Thus, higher fat consumption is likely to increase the overall energy intake. But this is also true of the consumption of sugars, such as high-fructose corn syrup, which contains 4 kcals per gram and may not give the feeling of "satiety", thereby leading to an increase in intake. The problems are compounded by the fact that dietary assessment in developed countries is complex; such issues are discussed in section 7.1. Further, measures of obesity such as waist and hip circumferences ("central obesity") are important risk factors for coronary heart disease. Section 7.1 briefly summarizes the relevant anthropometric assessment literature for the general reader. Also, the results from an analysis of data from Women's Health Trial: Feasibility Study in Minority Populations (WHTFSMP) for lowering fat intakes are summarized. The analysis showed that higher intakes of saturated, monounsaturated and polyunsaturated fats were not systematic predictors of anthropometric measures. Rather, women with "unhealthy" eating habits and low levels of physical exercise were heavier and had higher waist and hip circumferences. Implications of these findings for food policy are discussed.

Section 7.2 describes various approaches in the psychological literature to dietary modifications and obesity; alternative theories of human behavior influence strategies for promoting dietary changes in studies in developed countries. For example, the "health belief model" emphasizes the importance of individuals' perceptions of risks of contracting diseases, and the perceived risks may facilitate dietary changes. This approach contrasts with the utility maximization assumption in the economics literature. Ways of integrating the approaches are discussed. Further, psychologists have designed interventions for reducing child obesity in schools and certain studies are described in section 7.2. Last, because the questionnaire in WHTFSMP was designed by nutritionists, epidemiologists and psychologists, one can quantify relative magnitudes of psychological variables on changes in dietary intakes. The results reported in section 7.2 show that women's perception of health risks and "self-efficacy" were important predictors of dietary intakes, especially in the intervention group. Moreover, educated women made greater changes in the intervention group, though this was generally not the case for the control group. Because low-income populations in the US have poor dietary habits, it would be desirable to provide nutrition education, as is done in the Supplemental Food Program for Women, Infants and Children. While the costs of such programs are high, healthy diets and lifestyles can result in major savings in medical expenditures. While the long-run sustainability of dietary changes needs to be further investigated, the efficacy of nutrition education programs is likely to be enhanced by incorporating psychological variables.

Section 7.3 describes the approaches to obesity in the economics literature. Low food prices in grocery stores and fast-food restaurants, higher household incomes, and sedentary lifestyles are potentially contributing to weight gain among populations. Moreover, large numbers of foods available in grocery stores can complicate the conventional modeling of demand for food. The "characteristics" models for demand for food, emphasizing the importance of invoking assumptions that are consistent with knowledge in the biomedical sciences, are discussed. Furthermore, a study emphasizing the role of food processing technologies in increasing intakes is discussed (Cutler *et al.*, 2003). Also, articles relating maternal hours of employment to childhood obesity and a study linking low food prices to obesity in the US are discussed. This section also discusses strategies for reducing the prevalence of obesity, taking into account nutritional, psychological and economic approaches. In view of the high costs of caring for chronic diseases, an integrated approach is essential for stemming the obesity epidemic. While several strategies for obesity prevention are outlined, it is unlikely that major progress can be made without substantial resources devoted to educating the public about the benefits of healthy diets and lifestyles. Moreover, subsidies to large farmers and agribusiness in developed countries need to be reappraised, since they may adversely affect population health in developed and developing countries. Finally, Chapter 8 summarizes the main implications of the material presented in Chapters 2–7 and suggests future research that can facilitate development of effective food and health policies for improving well-being.

2

Demand for food and nutrients in developing countries

2.1 Introduction

The demand for food and nutrients in developing countries is a broad and important topic because it is useful to learn how dietary patterns change with incomes. For example, in rapidly growing economies, if diet quality, reflected in the intakes of micronutrients such as iron and vitamins A and C, improves with incomes, then policy-makers need not be too concerned with design of specific food policies to improve nutritional intakes. By contrast, if diets are low in iron and intakes do not increase with incomes, then policies of distributing iron tablets to vulnerable groups such as mothers and children will enhance their iron status, which is important for children's learning. However, researchers in fields such as economics, nutrition, anthropology and public health have adopted contrasting approaches to various problems. Most authors emphasize only a few aspects, whereas the use of a broad analytical framework is essential for understanding the conceptual and empirical issues and for formulating food policies.

It is generally accepted that food consumption depends on individuals' physiological needs and on cultural factors. Moreover, anthropologists are interested in cultural factors affecting food intakes, while nutritionists are typically more concerned with the effects of food intakes on health and disease outcomes. In contrast, economic analyses often approach food consumption based on models for intakes that assume individuals or households maximize a utility function subject to budget constraint. Thus, food intakes in the economics literature depend on food prices and household incomes. Of course, cultural factors are important and diets in different parts of the world are different. For example, many populations in South America consume tortillas and beans, whereas in south Asia, rice and wheat are staple foods that are consumed along with pulses (lentils) and vegetables. The food grown depends on soil and climate conditions so that dietary patterns have evolved gradually, taking account of production constraints. As an example of the influence of cultural factors on food consumption, one can think of countries such as India, where many people have been vegetarians for thousands of years and variation in meat prices has not affected their behavior. With increasing "globalization", however, attitudes are changing and meat consumption among formerly vegetarian households is on the rise.

The determination of individuals' nutritional needs is complex and such issues have been studied by nutritionists in population and laboratory settings. It is often the case that individuals eat according to their needs using approximate criteria. For example, if energy intake is insufficient for performing daily tasks, then the individual is likely to feel hungry and can satisfy his or her energy needs by consuming foods that are high in carbohydrates, protein and fats. By contrast, individuals may not realize their immediate requirements for "micronutrients" such as vitamins and minerals. For example, fruits and vegetables are good sources of vitamins A and C and if these intakes are low, eventually individuals are likely to develop conditions such as nutritional blindness, as in parts of Indonesia (Sommer, 1990). Moreover, human immune systems require regular intake of micronutrients, so that diets have gradually evolved to supply adequate quantities. However, with the recent availability of a variety of energy-dense foods at low prices, individuals may over-consume food, leading to weight gain; issues surrounding the increase in obesity will be discussed in Chapter 7.

Sources of dietary energy

Dietary energy is derived primarily from carbohydrates, fats and protein, which are often referred to as "macronutrients". Carbohydrates are found in plant-based foods such as wheat, rice and other staple foods consumed by populations. Human beings must also consume some fats, even though fat is energy-dense in that each gram contains 9 kilocalories (kcal) of energy; a gram of carbohydrate (and protein) typically contains 4 kcal of energy. Moreover, protein intake is essential for repairing wear and tear of muscle and a certain percentage of dietary energy must come from protein. Some of the amino acids are found mainly in protein from animal sources, so that the "quality" of the diet, especially in developing countries, is reflected in protein intakes from animal foods such as meat and milk. There are complex interactions between energy and protein intakes (Hutchinson, 1969). For example, protein consumed without adequate dietary energy is broken down by the body to meet energy needs. An understanding of these interactions is important for devising food policies. For example, suppose that there was a drought in a region and people were hungry and social scientists were asked to develop suitable food policies. It would be important first to eliminate energy deficiencies via staple foods so that one can be sure that protein from foods such as meat will not be utilized to meet energy needs.

The problems of nutrient interactions are also important from the standpoint of nutrient absorption, which is emphasized by nutritionists. For example, vitamin C intakes increase the absorption of minerals such as iron (Monsen and Balinfty, 1982), which is important for daily activities and for immunological functions (Scrimshaw and SanGiovanni, 1997). In the early economics literature Stigler (1945) noted that calcium from spinach is not absorbed, because of the presence of oxalic acid; it is important to incorporate nutritional knowledge in devising "optimal" diets, which are also influenced by socio-economic constraints faced by individuals. Further, iron deficiencies are prevalent in developing countries (UNICEF/WHO, 1999). Because iron absorption from grains is enhanced by larger quantities of vitamin C and meat, incorporating biomedical knowledge into food policy analysis can greatly enhance communication between policy makers trained in the fields of nutrition and economics. More importantly, under-nourished individuals in developing countries will benefit from efficacious food policies based on scientific knowledge.

Minerals and vitamins

The vitamin groups such as A, B and C, and minerals such as iron and calcium, are essential for maintaining health. Vitamin A is found in high concentration in fruits and vegetables and so is vitamin C; β-carotene is converted by the human body to vitamin A. Nutritionists underscore the desirability of consuming fruits and vegetables to ensure adequate intakes of vitamins A and C and other nutrients such as dietary fiber. Moreover, data on three subgroups of vitamin B, i.e. riboflavin, thiamine and niacin, are often analyzed and adequate intake of niacin is essential for preventing conditions such as pellagra. One of the properties of natural foods is that most nutrients are present in practically all foods, though in different proportions. For example, there is a small quantity of vitamin A in staple foods such as rice. However, green leafy vegetables are good sources of vitamins A and C. While one would expect diet quality to improve with incomes, the education and dietary knowledge of population subgroups even in developed countries can be poor (see Chapter 7). For example, approximately 40% of the dietary energy in the US is derived from fat and may contribute to chronic medical conditions such as cancers and heart disease. A low-fat diet consists of 20% of energy from fat as in countries such as Thailand, where prevalence rates of breast cancers are low (Keys, 1980). Thus, food choices may be adversely affected by poor nutritional knowledge, especially as economic prosperity leads to abandonment of traditional diets in favor of palatable energy-dense foods. Economists modeling food intake data need to be careful in invoking assumptions about dietary knowledge and behavior, so that their analyses appeal to a broad spectrum of scientists.

Cultural factors and diet

In empirical studies it is important to incorporate the influence of cultural factors affecting food consumption. In traditional societies, food consumption mainly depended on what could be grown; climate is still an important determinant of food consumption in low-income countries. For example, rice and wheat have emerged as staple foods over centuries and the choice between them is strongly influenced by production constraints. An important contribution of economists has been to recognize that cultural factors affect diet. For example, Gorman (1968) noted that "choices depend on tastes and tastes depend on past choices". Implicit in this phenomenon of "habit persistence" is that individuals tend to consume similar diets and changes occur gradually over time. For example, if the price of milk falls in a developing country, then parents may continue to offer milk to children for breakfast and may also increase milk consumption at other meals. The milk consumed will gradually reach an "optimal" level. From a modeling standpoint, "lagged" or previous level of demand should be an explanatory variable for modeling the current demand in "dynamic demand" models that reflect habit persistence. There are other features of foods for which they are desired, such as their mineral and vitamin "characteristics", and these can be incorporated to some extent in models for demand (Gorman, 1980). However, in an affluent country such as the US, changes in prices may not have dramatic effects on food consumption. By contrast, demand in developing countries for nutritious foods such as vegetables and milk may be very sensitive to price and income fluctuations, especially for the poor.

It is also interesting to compare the coefficients of explanatory variables such as income and education in models explaining food intakes in developing and developed countries. For example, in the analysis of data from the US, one would be more interested in the effects of "unhealthy" eating habits and low education levels, and nutrition education programs can attempt to change dietary behavior (see Chapter 7). By contrast, one is likely to find that individuals in developing countries may not be consuming adequate quantities of vitamins and minerals due to high food prices and low incomes. Policy-makers may need to devise policies that encourage production of foods such as meat and vegetables in developing countries. It is important to take into account the circumstances of individuals making food consumption decisions. This is sometimes overlooked in the economics literature because a "substantive" definition of rationality underlies models of utility maximization (Simon, 1986). In contrast, psychologists use a "procedural" definition of rationality which is helpful in the context of dietary behavior and change (Prochasks and De Clementi, 1984). For example, in the procedural approach, it is important to investigate if individuals behave in a consistent manner and if factors such as greater "self-efficacy" can facilitate dietary and lifestyle changes (see again Chapter 7).

2.2 Socio-economic determinants of food consumption in developing countries

Food intakes are likely to be influenced by food prices, household incomes, the education and knowledge of decision-makers, tastes, and individuals' energy requirements. While the role of prices and incomes has been emphasized in the economics literature, the design of surveys is important for determining the types of analyses that can be conducted. For example, if the survey covers households located in different geographical areas, then one might analyze the effects of food prices on consumption. By contrast, if the data are taken from a single geographic location, then food prices are likely to show small variation and one can investigate the effects of variations in household incomes and other variables on food intakes. Moreover, the energy and nutrient requirements of household members affect their food intakes. For example, the energy requirements of subsistence farmers are likely to exceed those of white-collar workers in skilled occupations. An individual's energy requirements depend on the basal metabolic rate (BMR), that is the minimum energy necessary for sustaining life. BMR in turn is influenced by the height and weight of individuals and also by their muscle and fat mass (James and Schofield, 1990; Johnstone *et al.*, 2005). Furthermore, BMR is important for predicting energy expended by individuals on various activities (FAO/WHO/UNU, 1985). For example, the energy expended in walking slowly is 2.8 × BMR, whereas the energy expenditure for cutting sugar cane is 6.5 × BMR. Thus, in order to interpret the effects of socio-economic and physiological variables in models for food intakes, investigators need to have an understanding of relevant issues in the nutritional sciences.

Many economists have estimated the effects of food prices on households' food consumption in developing countries. For example, Pitt (1983) and Pitt and Rosenzweig (1985) estimated the effects of food prices on households' intakes of dietary energy in Bangladesh and Indonesia, respectively. Because these households in Bangladesh and Indonesia were distributed across the country, analyses can pick up some of the effects of food prices on consumption. Even so, levels of economic activity can differ between regions and households' energy intakes are influenced by the occupations of their members, which in turn determine energy requirements. One might control for such factors using indicator variables for different geographical regions. In general, it is likely that cross-section regressions covering households from different regions reflect the "long-term" effects of food prices on food consumption (Chapter 6). If one wants to look at the effects of variability in economic factors ("Engel curves") in a short time frame, then longitudinal (panel) data on food intakes are useful. It should be noted that the nutrition literature no longer refers to energy as "calories", in part because energy is not a nutrient. Moreover, the human body uses the heat released by burning food for performing tasks, and energy intake is now expressed in kilo- or megaJoules (1 kcal = 4.18 kJ).

Policy issues

Policy-makers concerned with food consumption and health outcomes are interested in knowing how diets change with prices and incomes (e.g. Pinstrup-Andersen, 1988). If, for example, individuals nourish themselves better with rises in incomes, then policy-makers need not be too concerned, especially in regions enjoying high economic growth. Further insights into this issue can be gained via analyses of food intakes data from household surveys. However, one might distinguish between the situations where there are food shortages and where the quality of diet is poor due to protein and micronutrient deficiencies. For tackling food shortages ("energy deficiencies"), programs such as distributing subsidized or free food are essential (Drèze and Sen, 1990). Also, schemes such as "food for work" are useful especially if individuals are in good health and there are few employment opportunities. Moreover, price subsidies are often necessary to enable the poor to consume adequate quantities of energy and micronutrients. For example, the Indonesian government has subsidized rice for many years; one of the themes in the food policy literature is to analyse the effects of price subsidies on food consumption patterns (Timmer *et al.*, 1983). Changes in food consumption are often disaggregated into income and substitution effects due to price changes. For example, a fall in price of a food would lead to increased consumption, and so would an effective increase in income. Food policies have been formulated by economists taking into account the estimated magnitudes of these two effects.

Further, in order to interpret the effects of income on energy intakes, it is useful to have an understanding of the nutritional issues involved. For example, income elasticities of energy intakes based on household expenditures often fall into the range [0.5, 0.8] (see below). By contrast, using data on individuals' food intakes, estimated income elasticities are small and may fall within the range [0.0, 0.15]. It is true that aggregation reduces random variation in food intakes, so that elasticities based on aggregate food consumption data are likely to exceed those estimated from data on food intakes. However, if the income elasticity of energy intake is 0.8, then doubling of income implies an increase of 80% in energy intakes. While this may be plausible in areas with endemic hunger, it is physiologically impossible to increase energy intake by large amounts without concomitant increases in energy expenditures and/or body weight. Thus, while estimation of income elasticities from aggregate data is informative, one needs to explore the relationships using more direct estimates of food intakes for assessing the magnitudes. For example, it may not be correct to interpret small income elasticities estimated from food intakes data as meaning that individuals do not nourish themselves better with increases in incomes (Behrman and Deolalikar, 1987). In a similar vein, effects of price subsidies on food consumption should be reconsidered using individual-level food intakes data. A comprehensive approach to food policy formulation entails a broad understanding of various scientific and data issues.

Quality of diet

With increased food production, the common situation in developing countries is that quality of diet is poor due to protein and micronutrient deficiencies (IFPRI, 1996). Because individuals' nutrition requirements are complex, it is difficult to devise simple economic solutions for alleviating micronutrient deficiencies. Instead, as one incorporates nutritional complexities into economic models, differences in the approaches of economics and nutritional sciences begin to disappear. The "characteristics" model for demand for food was originally proposed in Gorman (1956) (see Gorman, 1980) and in this formulation, foods are demanded for their vitamin and mineral content. Lancaster (1971) and Ironmonger (1972) elaborated on some of these ideas. The idea that foods are demanded for their content seems rather obvious. For example, a hungry individual would prefer to satisfy energy needs by consuming foods such as carbohydrate and meat, which have a high energy content. In contrast, one would consume fruits and vegetables to satisfy the needs for vitamins. Because most nutrients are found in all foods, though in different proportions, one has to analyze demand for food and nutrients in a somewhat less direct but scientifically acceptable manner (see Chapter 7).

Stigler (1945) recognized many of the nutritional issues when calculating the "minimum cost of diet" using a linear programming approach. By contrast, Ironmonger (1972) applied the notion of "hierarchy in human wants" (Georgescu-Roegen, 1966) to illustrate demand for energy and nutrients. For example, it was argued by Ironmonger (1972) that given a limited amount of money, an individual would first satisfy the demand for energy, followed by the demand for protein and micronutrients. In practice, however, such a scheme is not feasible due to the energy and nutrient composition of foods. Rather, it is likely that a consumer will demand foods that are good sources of energy, protein and micronutrients in that order ("hierarchy"). For example, fruits and vegetables are good sources of micronutrients such as vitamins A and C but poor sources of dietary energy. Thus, relatively high demand for fruits and vegetables in comparison to staple foods is an indicator of higher demand for micronutrients. It is essential to use a flexible framework for assessing the effects of economic and other factors on diet composition.

Assessing the quality of diet from food intakes

Energy and nutrient intakes in social science research in developing countries are mainly assessed via food expenditures surveys. Expenditures data on food groups such as grains, meat, vegetables and milk are subsequently converted into energy and nutrients using food conversion tables such as those for India (Gopalan *et al.*, 1970). More recent research has measured food intakes directly via survey instruments such as the 24-hour recall method, which records all food intakes in the previous 24-hour period. Interviewers in developing countries visit households and ask members (or target individuals) "What did you consume in the last 24 hours?" and inquire about portion sizes. Food intakes are then converted into energy and nutrient intakes using a list of 200 or more foods. Alternative methods in the nutrition literature include "food records", where individuals fill out forms recording all food intakes for a period of three, four or seven days (see Chapter 7). While intakes data from food records covering several days are representative of individuals' "habitual" food intakes, such methods are not frequently used in developing countries because of the levels of proficiency required for completing the forms.

It would be helpful to discuss briefly the relative merits of using alternative methods for measuring food intakes in developing countries. Data from household expenditure surveys are usually in a very aggregated form and can provide useful insights since variation caused by factors such as food shortages is likely to be reduced. The variation for same individual over time is referred to as "within"-subject or "intra-individual" variation. However, actual food intakes are difficult to assess in data from food expenditure surveys since diet quality and food consumed by non-members ("guests") are difficult to measure. By contrast, the 24-hour recall method enables the collection of detailed information on foods consumed by each member in the household. Moreover, by visiting the household on a particular day, enumerators do not influence households' behavior. Such issues are regarded as critical in the nutrition and behavioral literature (Block, 1982) though economists have been less concerned with them. For example, while seven-day food records provide good estimates of nutrient intakes, such methods can influence individual behavior on the days that food records are filled in. By contrast, the 24-hour recall method does not influence behavior, though the day for which food intakes are recorded may not be "typical" and hence there may be large within-subject variation in intakes when longitudinal data are analyzed.

Further, no matter which dietary assessment method is used, it is useful to assess diet quality and investigate how it changes with incomes, education and nutrition knowledge in developing countries. One possible way of assessing diet quality is by expressing the intakes of protein and micronutrients such as vitamins and minerals as ratios to the total energy intake. For example, the ratio protein–energy intake will be higher for individuals consuming more meat. Similarly, the ratio vitamin C–energy intake will be higher for those consuming more fruits and leafy vegetables. These ratios are particularly useful in view of the fact that individuals' energy intakes are influenced by their energy expenditures, i.e. by physical activity levels. Moreover, one cannot measure energy expenditures without sophisticated methods such as the doubly labeled water method (Goldberg *et al.*, 1991). For example, an agricultural worker may need to consume 5,000 kcals per day to work in the fields. If this worker consumes very little meat, then protein intake may be around 50 g and the protein–energy ratio will be 0.01. By contrast, an affluent white-collar worker will have lower energy expenditure (say 2,000 kcals) and may consume 100 g of protein by consuming meat and beans. Thus, even in developing countries, one cannot assess the adequacy of diets by looking solely at energy intakes. In fact, the protein–energy ratio for the white-collar worker in the example is 0.05, that is five times as high as that of the agricultural worker. Thus, the implications of the characteristic model of demand and knowledge in the nutrition literature can be utilized to a certain extent by expressing intakes data as ratio to energy intakes. By modeling the ratios of energy intakes or adjusting for energy intake by including it as an explanatory variable in the models for nutrient intakes, one can analyze the effects of socio-economic variables on diet quality in a broad scientific framework.

2.3 Demand systems estimated for energy and nutrients using food expenditure surveys

As noted in section 2.2, an essential feature of the dynamic demand model is that demand for a commodity in the current period depends on the previous level, due to habit persistence. Because diets are changing with increased globalization, one might see less habit persistence, though it is reasonable to expect greater habit persistence in rural areas of developing countries. However, habit persistence may be less apparent as one disaggregates data on food groups or analyzes nutrient intakes, partly due to measurement issues discussed above. A dynamic demand function (q_t) can be written as a function of prices (p_t), income (y_t), and lagged quantity:

$$q_t = h(p_t, y_t, q_{t-1}) \qquad (2.1)$$

Other formulations for dynamic demand models are presented in Pollak (1968), Houthakkar and Taylor (1970) and Philips (1974). While the previous literature used aggregate time-series data for countries to estimate such models, economic and psychological theories pertain to individual behavior and hence modeling household demand is appealing. Moreover, even at the household level, equation (2.1) has been used for modeling quantities consumed on the basis of food expenditure surveys, i.e. for broad food groups. In this section, the findings from countries such as India and the Philippines will be presented for more disaggregated demand functions for nutrient intakes. The model in equation (2.1) can be enlarged to include household size, education level of decision-makers, and other variables. Note, however, that when longitudinal surveys are restricted to the same geographical region, there is not much variation in prices and researchers are primarily concerned with the effects of household incomes on food or nutrient quantities consumed.

There have been very few studies that estimate dynamic demand functions from longitudinal or panel data. Some of the methodological difficulties arise in the estimation of dynamic models when the number of individuals (n) is large but the number of time observations (t) is fixed. The methodological issues were addressed by Bhargava and Sargan (1983) by extending simultaneous equations estimators from the econometrics literature to longitudinal data. The reader need not be concerned with the technical details of maximum likelihood estimation, and the main implications of the diagnostic tests applied to models will be pointed out when interpreting empirical evidence. In the next section, estimates of income elasticities of six broad food groups are presented, using data from a survey conducted by the International Crops Research Institute for Semi-Arid Tropics (ICRISAT) in India.

Income elasticities of food groups estimated using ICRISAT data from India

ICRISAT–conducted a study of nutritional and economic factors in six villages of southern India in 1976/7 covering approximately 240 households (Binswanger and Jodha, 1978). Data were collected on demographic and economic variables and on anthropometric measures—height and weight—in four survey rounds that were separated by six-month intervals. Because some of the economic information was available on an annual basis, income elasticities of foods (and nutrients) were estimated using annual averages for 1976 and 1977 in three villages. The empirical model for consumption of grains (and sugar, pulses, vegetables, milk and meat) estimated by Bhargava (1991a) were:

$$
\begin{aligned}
\ln(\text{Grains})_{it} = {} & a_0 + a_1 \text{ (Village dummy 1)}_i + a_2 \text{ (Village dummy 2)}_i \\
& + a_3 \text{ (No. of adult equivalents)}_{it} \\
& + a_4 \text{ (No. of adult equivalents)}_{it}^2 \qquad\qquad (2.2) \\
& + a_5 \ln(\text{Total household expenditure})_{it} \\
& + a_6 \ln(\text{Grains})_{it-1} \\
& + u_{it} \; (i = 1, \ldots, N; \, t = 1,2)
\end{aligned}
$$

In the model in equation (2.2), "ln" represents natural logarithms and this transformation is useful for reducing "noise" or variation in the data. Two indicator variables were included for the villages. The number of adult equivalents was constructed taking into account children's ages and gender differences in energy requirements. The empirical model in (2.2) was non-linear in the number of adult equivalents. Total household expenditure was a proxy for household income and can be potentially correlated with the error terms (u_{it}) (see below). Previous consumption of grains was included as an explanatory variable. Thus, the short-run income elasticity (percentage change in a dependent variable resulting from a 1% change in an explanatory variable) of grain was a_5, while the long-run income elasticity was $[a_5/(1-a_6)]$. The error terms (u_{it}) are often decomposed in a simple random effects fashion as

$$
u_{it} = \delta_i + v_{it} \qquad\qquad (2.3)
$$

where δ's are individual specific random variables that are distributed with zero mean and constant variance, and v's are independently distributed random variables with zero mean and constant variance. It is important to include individual effects (δ_i) in longitudinal data analysis since these reflect unobserved individual characteristics ("heterogeneity"). Ignoring such features can lead to spurious results since some relevant variables are likely to be omitted from econometric models. Further, in the empirical modeling, a more general formulation was used where it was assumed that the u_{it}'s were distributed as multivariate normal variables. While the random effects decomposition in equation (2.3) was a special case of the multivariate normality assumption, with only two time observations, the two formulations are mathematically equivalent. The initial observations on grains consumption were treated as "endogenous" variables (correlated with the errors) in the instrumental variables and maximum likelihood estimation. The exogenous variables in the model were used as instruments and time-varying variables were treated as different in different time periods, which is useful for achieving identification of model parameters. Moreover, correlation between random effects and Total household expenditures was tested using likelihood ratio tests. If the exogeneity null hypothesis was rejected, then the models were estimated treating Total household expenditure as an endogenous variable.

Table 2.1 Maximum-likelihood estimates of household expenditure elasticities of foods using ICRISAT data

Variable	Model for food group					
	Grains[a]	Sugar	Pulses	Vegetables	Milk	Meat[b]
Constant	0.97	2.27	36.45	25.68*	−66.37*	−5.83
	(0.61)	(5.88)	(25.45)	(6.0)	(2.55)	(5.18)
Dummy 1	0.67*	−24.09*	−59.26*	−22.92*	−0.50	5.87*
	(0.09)	(4.27)	(20.82)	(3.40)	(3.42)	(2.39)
Dummy 2	0.30*	−7.54*	−9.28*	−4.28	−4.45	0.21
	(0.08)	(2.68)	(5.39)	(2.80)	(3.19)	(2.35)
Number of adult equivalents	−0.23	3.17*	−1.22	−1.94*	12.10*	0.36
	(0.13)	(1.16)	(1.72)	(0.56)	(0.77)	(0.51)
(Number of adult equivalents)[a]	0.06	−0.24*			−0.74*	
	(0.05)	(0.08)			(0.07)	
Expenditure	0.841*	0.035*	0.062*	0.056*	0.062*	0.016*
	(0.058)	(0.007)	(0.012)	(0.007)	(0.007)	(0.006)
Lagged dependent variable α	−0.197*	0.347*	−1.075	0.027	0.927*	0.612*
	(0.111)	(0.098)	(0.680)	(0.098)	(0.128)	(0.120)
Short-run elasticity[c]	0.841	0.640	1.674	0.813	1.094	0.579
Long-run elasticity[d]	0.703	0.980	1.674	0.813	14.980	1.492
Chi-square (2)[e]	0.543	1.452	0.328	0.241	0.238	5.033
2L[f]	453.7	−905.2	−1100.8	−850.8	−1223.1	−763.7

Note: 94 households were surveyed over in 2 time periods; asymptotic standard errors in parentheses.
[a] Sum of expenditures on rice, wheat, *jowar*, *bajra* and maize in logarithms.
[b] Vegetarians excluded.
[c] Elasticities calculated at sample means.
[d] Long-run elasticity = elasticity/$(1 - \alpha)$.
[e] Chi-square test for exogeneity of Total household expenditure.
[f] Twice the maximized value of the log-likelihood function.
* $P < 0.05$
Source: Bhargava (1991b).

The results from estimating the model in equation (2.2) by maximum likelihood for grains, sugar, pulses, vegetables and meat are reproduced from Bhargava (1991b) in Table 2.1. First, coefficients of Total household expenditure were invariably significant at the 5% level in all six models. Second, the long-run elasticities of sugar, meat and especially milk are all higher than their short-run counterparts. For example, the lagged dependent variable was estimated with the coefficient 0.97 in the model for milk, which led to the very high long-run income elasticity of milk. The short-run income elasticity of milk was 1.09, while the long-run elasticity was almost 15. Milk is an important source of protein, calcium and other nutrients and its consumption was likely to be influenced by household incomes, especially in the long run. Third, the long-run elasticity of grains was lower than the short-run elasticity since the lagged dependent variable was estimated with a negative coefficient. This was also found in previous studies of aggregate demand in the US (Houthakkar and Taylor, 1970), especially for commodities that may be viewed as "inferior". Fourth, the relationships between food groups and household size expressed in terms of number of adult equivalents were quadratic in the models for grains, sugar and milk. Other background characteristics of households, such as the education level of the head, were not statistically significant in these models. Last, the chi-square tests for exogeneity of Total household expenditure were not statistically significant. Insignificance of likelihood ratio tests may in part be due to the large variation in consumption patterns, which can widen the confidence intervals for the test statistics.

The empirical results presented for Indian households seem plausible from many viewpoints. While one might expect long-run income elasticities of nutritious foods to be larger than their short-run counterparts, this is unlikely to be the case for staple foods such as grains. In some of the models, the lagged dependent variable was not significant presumably because only two time observations were available. It is worth noting that meat consumption in India is critically influenced by religious factors, and vegetarian households are unlikely to change their dietary behavior with increases in incomes. In fact, the results in Table 2.1 dropped 12 vegetarian households when estimating the model for meat consumption. Moreover, for meat-consuming households, short-run income elasticity was 0.58, whereas the long-run income elasticity was 1.5. Thus, there were greater differences between short-and long-run effects, especially for nutritious foods such as milk and meat, which are good sources of protein, energy, calcium and iron. The income elasticities of energy and nutrients estimated from food intakes data are discussed in section 2.4. As discussed above, these are likely to be smaller in magnitude than income elasticities estimated using aggregate food expenditures data.

Income elasticities of energy intakes using data from food expenditure surveys

Early research in economics compiled food expenditures data on households and converted the "intakes" of various food groups into energy and nutrients using food conversion tables. The income elasticities of energy and nutrients were estimated using econometric models for cross-sectional data; effects of food prices were investigated in applications where the data were nationally representative. Moreover, it is important to control household size in the models though it is generally infeasible to control for members' energy requirements, since they depend on individuals' height, weight and daily activities. The empirical models for energy intakes provide a useful starting point for examining the relationships between household incomes and energy intakes; empirical evidence is important in the formulation of food policies. Following the literature, a simple model for estimating income elasticities of energy can be written as:

$$\ln(\text{Energy intake})_i = b_0 + b_1 \ln(\text{Household income}) + b_2' X + u_i \quad (2.4)$$

While household income is included in equation (2.4), total household expenditures are often substituted for incomes, owing to the difficulties in imputing incomes in developing countries. The matrix X consists of explanatory variables such as household size, maternal education, and the prices of various food groups. Income elasticity of energy intakes (b_1 in equation (2.4)) was estimated to be around 0.80 for Bangladesh by Pitt (1983) and was estimated at 0.60 for Sri Lanka (Sahn, 1988). Moreover, Alderman (1986) estimated the income elasticity of energy intakes for India to be 0.45, while it was 0.15 for Indonesia (Ravallion, 1990), which seemed lower than for other countries.

Furthermore, since income elasticities estimated from food expenditures data were typically large, policy-makers expect to see large increases in food intakes with rises in household incomes. However, there may be non-linearities with respect to incomes, so that income elasticities may be large only for very poor households. Moreover, economic analyses based on income and substitution effects of changes in food prices were viewed as providing critical input for food policy analysis (Timmer *et al.*, 1983). From a nutritional standpoint, unless there was widespread hunger, one would not expect to find large differences in energy intakes between poor and well-off households though sources of dietary energy will differ. For example, the poor may derive the bulk of their energy by consuming staple foods such as rice or wheat. In the absence of widespread hunger, one might view income elasticities in the neighborhood of 0.50 to be large, reflecting major differences in the quantity of food consumed by the rich and the poor. Moreover, such estimates are likely to reflect high levels of unemployment among the poor, which depress energy expenditures and hence food intakes. Even in such situations, the poor cannot go hungry for extended periods and one is likely to observe lower body weights, i.e. widespread "wasting" (such as low weight-for-age among children). This normally happens when there are food shortages due to droughts and/or wars. Also, micronutrient deficiencies can increase morbidity and contribute to wasting.

There are several explanations for the large income elasticities of energy intakes estimated using food expenditures data from developing countries. First, because the data are in an aggregate form and food expenditures are a high proportion of total expenditures, there is likely to be a strong correlation between food budget shares and annual income or expenditures. Moreover, diet quality is likely to improve with the perceived affluence of households. For example, maize may be replaced by rice as a staple as household incomes increase. Even in situations where rice is a staple food, the poor are likely to consume cheaper varieties of rice; such varieties may have a lower micronutrient content, such as less iron, and may not taste as good. Such issues cannot be addressed using aggregate food expenditures data. While the magnitudes of income elasticities is revisited in section 2.4, another reason for high income elasticities based on food expenditures data is that there are food "leakages" due to workers eating with landowning and well-off households (Bouis and Haddad, 1992). This phenomenon increases food consumption in better-off households and this may also be true for urban households that offer food to workers performing household tasks. Thus, it is important to estimate income elasticities of energy and nutrients using data on food intakes, though such analyses will present other complications.

2.4 Income elasticities of energy and nutrient intakes using data on food intakes

As discussed in section 2.2, 24-hour recall surveys of food intakes can be easily implemented in developing countries to provide estimates of food intakes. The data on (say) 200 food groups recorded is converted to energy and nutrient intakes using food composition tables such as those of Gopalan *et al.* (1970) for India. Moreover, the "International Minilist" recorded 1,800 foods in Egypt, Kenya, Mexico, Senegal, India and Indonesia (Calloway *et al.*, 1994) and food intakes can be converted to energy and nutrient intakes. Because 24-hour recall surveys were conducted in countries such as India, the Philippines and Bangladesh, income elasticities of energy and nutrient intakes for these countries are discussed in section 2.5 and certain conceptual and methodological issues are addressed.

Behrman and Deolalikar (1987) analyzed the ICRISAT data by aggregating nutrient intakes for household members using the ordinary least squares method, which is appropriate for cross-section regressions. The main findings were that income elasticities of energy and nutrients such as protein, calcium, iron, vitamins A, B (niacin, riboflavin and thiamine) and C were not significantly different from zero. Thus, the authors concluded that individuals do not nourish themselves better with increases in incomes and may be consuming foods because of "taste, odor, smell, and status value". However, these conjectures seem implausible because "status value" foods are seldom available in villages in the ICRISAT surveys. Rather, food consumption patterns are stable and there is likely to be habit persistence in diets. For example, the poorest households typically consume rice and, as their incomes go up, they might add dishes containing pulses, vegetables and meat, depending on the prices. By contrast, better-off households are likely to consume pulses, vegetables, milk and meat on a regular basis. Thus, income elasticities, especially of micronutrients, should be positive though perhaps small in magnitude.

One of the problems with the models estimated by Behrman and Deolalikar (1987) was that the individual intakes data were aggregated for household members using recommended dietary allowances (RDAs) of energy and nutrients, which depend primarily on members' ages. Because the RDAs are based on several approximations and include a wide margin of error, aggregation based on RDA can induce further noise into the intakes data. This is in addition to the already high within-subject (intra-individual) variation in 24-hour recall data; food intakes on a randomly selected day may not reflect the "habitual" intakes, especially if households are poor and spread the consumption of nutritious foods. An alternative approach would be to estimate income elasticities of energy and nutrients using longitudinal random-effects models for individuals' intakes. Moreover, one can control for individuals' height and weight in the models since these variables are proxies for energy requirements. Habit persistence in diets can also be incorporated in the models, as discussed in section 2.5. First, the estimates of income elasticities from a static model for intakes are reported (Bhargava, 1991a):

$$\ln(\text{Nutrient intake})_{it} = b_0 + b_1 \ (\text{Village dummy 1})_i$$
$$+ b_2 \ (\text{Village dummy 2})_i$$
$$+ b_3 \ \ln(\text{No. of adult equivalents})_{it}$$
$$+ b_4 \ \ln(\text{No. of adult equivalents})_{it}^2 \qquad (2.5)$$
$$+ b_5 \ \ln(\text{Household income})_{it}$$
$$+ b_6 \ \ln(\text{Height})_{it} + b_7 \ \ln(\text{Weight})_{it}$$
$$+ u_{it} \ (i = 1, \ldots, 364; \ t = 1,2)$$

The model in equation (2.5) does not involve aggregation over household members and partly accounts for individuals' energy requirements. However, individuals' body weight may be correlated with the random effects (δ_i) in the error terms (u_{it}), especially in the model for energy intakes. For example, individuals may have certain unobserved characteristics, such as tastes for certain types of energy-dense foods, that increase their energy intakes and body weights. Thus, in the model for energy intakes, weight can be correlated with individual specific random effects (δ_i). One can use a stepwise "instrumental variable" estimator such as 3 stage least squares for consistent and efficient estimation of the model parameters under the assumption that body weight was correlated with the random effects; more general assumptions on the correlation pattern between explanatory variables and the error terms (u_{it}) were also invoked in the analysis. Instrumental variables estimation is facilitated by longitudinal data because realizations of time-varying variables in different time periods provide additional instruments that aid identification and estimation of the model parameters.

The estimated income elasticities of energy, protein, calcium, vitamins A and C, and vitamin B for the ICRISAT data using the model in equation (2.5) were 0.05, 0.06, 0.10, 0.10 and 0.13, respectively. These elasticities were significantly different from zero except for the income elasticity of iron intakes. As discussed in section 2.5, iron intakes are difficult to model because there are large discrepancies between total iron intake and the intake of "absorbable" (or bioavailable) iron (Bhargava, Bouis *et al.*, 2001). The results from static models for energy and nutrient intakes show that households nourish themselves better with increases in incomes and that it is reasonable to expect small income elasticities of energy and nutrients.

Of course, incomes may increase very slowly in certain developing countries and it may be necessary for national and international agencies to devise food policies, such as those subsidizing foods that are good sources of protein, vitamins and minerals, to ensure balanced nutrition, especially for children. The effects of nutrient intakes on children's health status and cognitive development will be addressed in Chapters 3 and 4, respectively.

2.5 Income elasticities of nutrient intakes in a dynamic framework

The static model in equation (2.5) does not incorporate habit persistence in diets and also does not address issues of diet quality that arise partly because nutrients are found in all foods though in different proportions. Some of these issues are addressed using the data from the Philippines, which cover a greater number of individuals observed in four time periods.

Table 2.2 Maximum likelihood estimates of income elasticities of nutrients in the simple dynamic models using ICRISAT data

Variable	Models for energy and nutrients				
	Energy	Protein	Calcium and iron	Vitamins A and C	Vitamin B
Constant	−0.646	1.362	10.617*	0.114	1.208
	(3.230)	(1.821)	(5.241)	(0.362)	(1.928)
Dummy 1	−0.007	−0.069	0.794*	−0.207	−0.230*
	(0.093)	(0.149)	(0.379)	(0.126)	(0.082)
Dummy 2	−0.274*	−0.293*	0.641*	0.071	−0.283*
	(0.051)	(0.048)	(0.323)	(0.079)	(0.049)
Number of adult equivalents	−0.064	−0.164*	0.052	−0.365*	−0.169*
	(0.064)	(0.049)	(0.119)	(0.065)	(0.051)
Household income	0.066*	0.106*	0.007^a	0.095*	0.127*
	(0.032)	(0.034)	(0.048)	(0.047)	(0.027)
Weight	−0.169	0.098	0.109	0.113	0.203*
	(0.332)	(0.165)	(0.131)	(0.155)	(0.051)
Height	−0.159	−0.391	−3.609	0.198*	−0.119
	(0.518)	(0.476)	(1.744)	(0.048)	(0.437)
Lagged dependent variable α	1.409*	0.947*	2.387*	0.454	0.582*
	(0.608)	(0.358)	(0.790)	(0.330)	(0.205)
2L	1937.9	1793.9	1482.2	730.0*	1496.2

Note: All variables are in natural logarithms; 364 individuals were observed in two time periods; asymptotic standard errors in parentheses.
^{a} Significant with inclusion of square of household income.
* P < 0.05.
Source: Bharghava (1991b).

35

For the ICRISAT data, Bhargava (1991b) estimated a simple dynamic model for energy and nutrient intakes by including lagged (previous) intakes in the model in equation (2.5). The findings are briefly summarized here and the reader can refer to this article for details.

In the simple dynamic model, the income elasticities of energy, protein, vitamins A and C, and vitamin B were estimated to be 0.07, 0.11, 0.10 and 0.13, respectively (Table 2.2). All the income elasticities were significant at the 5% level. However, coefficients of the lagged dependent variables in the models for energy, protein, and calcium and iron were very large, indicating some type of model misspecification. These results could be due in part to the fact that the models were estimated using only two time observations. However, when nutrient intakes were expressed as ratios to energy intakes, income elasticities of protein–energy and vitamin B–energy ratios were statistically significant. Moreover, the lagged dependent variables in the models for protein–energy, vitamins A and C–energy, and vitamin B–energy ratios were 0.62, 0.13 and 0.51, respectively (Bhargava, 1991b: table 3). These results seemed plausible from a habit persistence standpoint. As argued in section 2.3, the protein–energy ratio is a good indicator of diet quality. Moreover, in terms of energy intakes, poor households are likely to consume a lower proportion of vitamins from fruits and vegetables than well-off households. By incorporating the nutritional aspects in econometric modeling, the empirical results from models for nutrient–energy ratios were an improvement over the results from simple dynamic models of energy and nutrient intakes. Moreover, the present approach contrasts with research in economics by Ironmonger (1972) and others that did not integrate nutritional aspects in the models and relied on assumptions regarding the nutrient composition of foods that are at odds with the biomedical knowledge.

Further, transformation of nutrient intakes into nutrient–energy ratios recognizes the importance of adjusting for individuals' overall energy intake, emphasized in the nutritional epidemiology literature (e.g. Willett, 1990). Because most nutrients are found in all foods, one cannot increase the intake of a nutrient without increasing intakes of several other nutrients. Thus, in econometric terms, the "structural" dynamic model for energy and nutrient intakes is a quasi-"reduced form" in the sense that other nutrients have been excluded from the set of explanatory variables. Using the ICRISAT data, Bhargava (1991b) investigated the effects of household incomes and other factors on nutrient intakes, while holding constant quantity of other nutrients consumed. For example, in the model for protein intakes, intakes of energy, calcium and iron, vitamins A and C, and vitamin B were included as explanatory variables in an extended version of the simple dynamic model. The technical issues in estimating the "interdependent" system are discussed in Bhargava (1991b). By solving the interdependent model, the income elasticities of energy, protein, calcium and iron, vitamins A and C, and vitamin B were 0.13, 0.18, 0.13, 0.18 and 0.31, respectively. These were higher than the corresponding results from static models in section 2.4, though the standard errors were also larger.

Lastly, there was some evidence in the results from the interdependent system on the issue of "hierarchy in human nutritional wants"; the likelihood ratio tests for exogeneity of energy and nutrients in the models for vitamins A and C, and vitamin B intakes, accepted the null hypotheses. Thus, decisions to consume foods that are high in vitamins A, B and C were likely to succeed the decisions to consume energy and protein. This treatment of hierarchy in nutritional wants was different from previous formulations (Ironmonger, 1972) and was consistent with the emphasis on cultural factors affecting diet. Also, nutritionists emphasize absorption of nutrients and nutrient absorption depends on the presence of various nutrients in the meal. By integrating the social science and biomedical issues in the analytical framework, empirical evidence showed the beneficial effects of increases in household incomes on intakes of desirable nutrients in India. Evidence on these issues is next summarized using data from the Philippines, Kenya and Bangladesh that have certain advantages over the ICRISAT data from India.

Income elasticities of nutrients in the Philippines, Kenya and Bangladesh

There were only two time observations available in the ICRISAT data for modeling the proximate determinants of energy and nutrient intakes. The Filipino survey followed 450 households in the Bukidnon region at four-month intervals for four time periods in 1984/5, and compiled extensive economic and nutritional information (Bouis and Haddad, 1992). The surveys in Bangladesh used a similar study design by following 900 households in the Saturia, Jessore and Mymesingh regions for three time periods in 1996/7 (Bouis, Briere *et al.*, 1998). A longitudinal survey in 1984/5 in the Embu region of Kenya was designed by psychologists and nutritionists (Neumann *et al.*, 1992); the income elasticities of energy and nutrient intakes by 100 school-aged children observed in three time periods at three-month intervals are presented (Bhargava and Fox-Kean, 2003).

The empirical models for intakes of energy, protein, calcium, iron, β-carotene, ascorbic acid, riboflavin, thiamine and niacin for 312 Filipino children observed in four time periods were estimated by Bhargava (1994). The models controlled for background variables such as mothers' housework and for children's height and weight. The short-run income (expenditure) elasticities of energy, protein, calcium, iron, β-carotene, ascorbic acid, riboflavin, thiamine and niacin were 0.08, 0.13, 0.23, 0.20, 0.21, 0.02, 0.24, 0.33 and 0.36, respectively. All the income elasticities were statistically significant except for ascorbic acid intakes, which are known to exhibit large within-subject variation, partly due to seasonal fluctuations in fruit and vegetable prices. The long-run elasticities were slightly larger because coefficients of the lagged intakes were typically close to 0.10. The estimated coefficients of lagged dependent variables were plausible given the level of disaggregation in recording data on food intakes and indicated some habit persistence in diets. The income elasticities for Filipino households were generally larger than those estimated using the ICRISAT data and were more precisely estimated due to larger sample sizes. For example, the income elasticity of energy and protein intakes from dynamic models for ICRISAT data were 0.07 and 0.11, respectively; corresponding income elasticities for Filipino data were 0.08 and 0.13, respectively, which were close. Income elasticities of micronutrients such as calcium, iron and vitamin B were higher for the Filipino sample, indicating greater improvements in the diet quality of children with rises in household incomes.

The data from Kenya contained relatively less information on income and expenditure variables; the "economic" index was based on measures of households' socio-economic status and cash income, which was recorded on a scale of 1 to 5. However, food intakes in the Kenya data were compiled very accurately; food intakes on two consecutive days were available every month, i.e. six days of food intakes were averaged to produce "habitual" intakes for the three-month period. The income elasticities, estimated by Bhargava and Fox-Kean (2003) using dynamic models for the intakes of energy, protein, calcium, iron, niacin, riboflavin, thiamine, vitamin A, and vitamin C, were 0.29, 0.39, 0.42, 0.31, 0.30, 0.31, 0.34, 0.26 and 0.37, respectively. Only the income elasticity of vitamin A intakes was not statistically different from zero. Income elasticities were generally higher in Kenya than in India and the Philippines. This was not surprising since there was a drought in the Embu region in 1984/5 and energy deficiencies were apparent (Cohen and Lewis, 1987). The effects of drought were also reflected in the large income elasticity of energy (0.39).

Finally, iron intakes are important because iron deficiencies and anemia are widely prevalent in low-and middle-income countries (UNICEF/WHO, 1999). Unlike many other nutrients, iron intake from staple foods such as rice and wheat are high but the absorption rates may be as low as 1–5% due to the presence of phytates in the meals. By contrast, iron from animal sources such as meat, fish and poultry is more readily absorbed (Monsen and Balintfy, 1982). For example, average iron intakes of women in the data from Bangladesh were 9.96 mg, while "absorbable" iron intake was only 0.85 mg (Bhargava, Bouis *et al.*, 2001). A woman's requirement of absorbable iron is at least 1 mg per day and hence Bangladeshi women were not meeting their requirement in spite of seemingly high iron intakes. From the standpoint of diet quality, it is of interest to estimate income elasticities of iron from all animal sources and from meat, fish and poultry. Income elasticities for these two subgroups of iron intakes were 0.65 and 0.71, respectively, and were highly statistically significant. Thus, diets of Bangladeshi women improved markedly with increases in household incomes; such information is useful for policy-makers in view of the fact that more than 50% of Bangladeshi women are anemic. It is evident that incorporating knowledge from the nutritional sciences in econometric models is useful for devising food policies for reducing nutrient deficiencies. For example, interventions in Bangladesh such as introducing new types of fish in ponds (Bouis, Briese *et al.*, 1998) can increase the intakes of "heme" iron, while improving seed varieties can increase vitamin A and C intakes, which are important enhancers of the absorption of "non-heme" iron from staple foods. Development of effective food policies is feasible using a multi-disciplinary approach and can reduce the prevalence of iron deficiencies in Bangladesh and other developing countries.

2.6 Conclusion

This chapter addressed various issues arising in the estimation of income elasticities of energy and nutrients using different types of data from developing countries. While it is reasonable to expect that food consumption depends on habit persistence and individuals' requirements, most microeconomic studies have not addressed these issues in detail. Moreover, from a food policy standpoint, it is important to investigate how diets change with household incomes in developing countries. Because food expenditures data are in an aggregate form, income elasticities of energy and nutrients are likely to be overestimated in such analyses. Thus, households' intakes of energy and nutrients are likely to increase at a slower rate than that predicted by income elasticities estimated using food expenditures data.

It is important for policy formulation to focus on energy and nutrient intakes using data on food intakes by individuals. While 24-hour recall data exhibit large within-subject variation, they provide useful information on food intakes, especially in developing countries. The empirical results from India, the Philippines, Kenya and Bangladesh presented in this chapter showed income elasticities of energy in the neighborhood of 0.10 unless there were food shortages, as in Kenya, where this elasticity was 0.39. By contrast, income elasticities of micronutrients ranged from 0.02 to 0.42. While high estimates of income elasticities of micronutrients are desirable because economic development will improve nutrient intakes, low elasticities indicate the need for interventions in the form of food subsidies to households to ensure adequate intakes. This is especially important for vitamin A and C intakes, which are strongly influenced by seasonal variations in fruit and vegetable prices, and for iron, whose absorption depends on nutrients such as vitamins A and C present in the meal. Adequate intakes of protein, iron, calcium and vitamins are important for maintaining health and for children's growth and learning, as discussed in Chapters 3 and 4.

From a food policy perspective, perhaps too much attention was given in previous economics research to energy deficiencies though, of course, they need to be tackled first. By incorporating knowledge from the biomedical sciences into econometric models for energy and nutrient intakes, researchers can formulate efficacious food policies for improving the nutritional and health status of inhabitants of developing countries. Income elasticities of nutrients presented in this chapter provide guidance to policy-makers for designing interventions that can be effective in reducing the prevalence of nutrient deficiencies, especially when nutrient intakes appear to respond slowly to increases in household incomes. Iron and vitamins A and C are particularly important nutrients for population health and policy-makers need to ensure higher intakes of dairy products and fresh fruits and vegetables in developing countries.

3

Nutrition and child health outcomes in developing countries

3.1 Background issues

In this chapter, the effects of socio-economic variables and nutritional intakes on indicators of child health such as height, weight and sicknesses will be investigated using data from developing countries. The analytical framework is sufficiently general and can be extended for analysing health outcomes in developed countries. It was emphasized in Chapter 2 that nutritional intakes (i.e. diet quality) improve with rises in household incomes in countries such as India, Bangladesh, Kenya and the Philippines. It is important to investigate the effects of energy and nutrient intakes on indicators of child health. For example, if higher intakes of vitamins A and C predict lower child morbidity, then policy-makers should support policies that make such foods affordable for poor households in developing countries.

Further, children in developing countries face harmful environmental factors, such as lack of sanitation, that can cause sicknesses and may lead to growth retardation. Growth retardation is reflected in low values of children's height-for-age ("stunting") and weight-for-age ("wasting") (Waterlow, 1994). Moreover, it is popular in the nutritional literature to use "z-scores" for height and weight based on a reference population such as the US (Hamill *et al.*, 1977) or the UK (Freeman *et al.*, 1995). For example, the z-score for a child's height can be written as:

$$z_i = (h_i - h_R)/s_R \quad (i = 1, \ldots, N) \tag{3.1}$$

where h_i is the height of the i'th child, h_R is the height in the reference population given age, and s_R is the standard deviation of height in the reference population for the child's age. More complex formulae are often used for constructing z-scores for body mass index (BMI), taking into account the characteristics of the distribution such as higher-order moments (Center for Disease Control, 2005). While certain methodological issues complicating the use of z-scores are addressed in Chapter 5, it is important to develop a conceptual framework for modeling anthropometric measures such as height, weight and sickness spells that is consistent with the biomedical literature. Moreover, children's arm circumference is a useful indicator of fat-free (muscle) mass (see e.g. Gibson, 1993), and it is important to develop an analytical framework reflecting the relationships between different anthropometric measures. Because dietary intakes play an important role in promoting children's physical growth and maintaining health, the effects of nutritional intakes on health outcomes are discussed next.

Dietary intakes and health indicators

Increased food production and availability in developing countries have greatly reduced energy deficiencies among populations. However, the diet quality of children is often poor in that intake of nutrients such as protein, iron, calcium, and vitamins A, B and C are below the optimal levels. Of course, adequate energy intakes are essential for maintaining daily activities and for children's growth. Since a high proportion of energy intake can be cheaply derived from staple foods, one would expect that indicators such as height are more strongly influenced by diet quality. This has important implications for combining children's height and weight as the weight-for-height or BMI, as is often done in the biomedical and social sciences. For example, if a child's linear growth is reduced due to inadequate intakes of calcium and other nutrients, then shorter children will appear to be well nourished if weight-for-height or BMI are used as indicators of nutritional status. As discussed in section 3.4, it is better to treat height and weight as separate health indicators (Kronmal, 1993), and apply a statistical test for whether they should be combined as weight-for-height or BMI (Bhargava, 1994). This approach is also useful for studying problems of overnutrition in developed countries, where a high BMI is a risk factor for chronic diseases such as diabetes and coronary heart disease (see Chapter 7).

Further, it is known in the nutritional sciences that specific nutrients play important roles in maintaining various health functions. For example, regular intakes of vitamins A and C are essential for enhancing the body's immune systems (Scrimshaw and SanGiovanni, 1997). Similarly, good iron status, reflected in blood hemoglobin and ferritin concentration, is important for physical activity and can reduce sicknesses (Chandra, 1977). Because body stores of such micronutrients are low, it is important to investigate the effects of dietary intakes on sicknesses. However, it is important to develop quantitative measures of sicknesses that can present methodological challenges. For example, a binary (0/1) outcome for whether a child was sick with diarrhea in the past month is difficult to model unless the sample size is very large (Cebu Study Team, 1992). Moreover, children in developing countries are often sick with multiple symptoms, and the intensity and duration of illnesses are deeper measures. For example, one can construct an index of morbidity that is based on the duration and intensity of sicknesses (e.g. Rand Corporation, 1983). The index sums the number of days for which a child is sick with different symptoms, with each symptom counting as a separate illness. It is also possible to assign different weights to different symptoms based on the severity of illnesses. In the example of gastrointestinal diseases, serious diseases such as typhoid and cholera entail lengthy sickness spells. Thus, even simple morbidity indices assigning equal weights to different symptoms will assume higher values due to long convalescence periods if children are sick with typhoid or cholera. Overall, the scores on morbidity indices reflect the length and intensity of illnesses and can facilitate the modeling of the effects of diet and environmental factors on child morbidity (Bhargava, 1994). The effects of environmental factors on child morbidity and how morbidity in turn can diminish children's physical growth are briefly discussed in section 3.4.

The biomedical approach versus economic approaches to modeling health indicators

The biomedical approach to modeling indicators of health is often driven by specific hypotheses regarding the effects of nutrients on health outcomes. For example, the US Agency for International Development sponsored three similar surveys in Egypt, Kenya and Mexico in 1984/5 to investigate the effects of energy derived from animal products on the cognitive development of children (Neumann *et al.*, 1992). While children consuming better diets may perform better on cognitive tests, there are several intermediate steps underlying cognitive development. For example, better-quality diets can enhance children's anthropometric indicators, such as height and weight, and can reduce sicknesses, thereby increasing school participation. If the school environment is stimulating, then better-nourished children may achieve higher scores on cognitive tests through these pathways (Bhargava, Jukes *et al.*, 2005). While it is recognized in the biomedical sciences that the development of the human brain is a very complex process, researchers often estimate simple correlations or associations between dietary intakes and cognitive measures using the data from their studies. In some situations, moreover, randomized trials are designed to assess the effects of changes in specific variables on health outcomes. Chapter 4 discusses how the data from randomized trials can be analysed to model the pathways underlying children's scores on cognitive tests in developing countries.

In contrast with the biomedical approach, economists assume the existence of a "health production function" that treats various socio-economic variables as "inputs", and variables such as children's height and test scores as "outputs", in the spirit of models for the production of commodities. Moreover, because households are assumed to have solved an "optimization" problem, health outcomes depend primarily on food prices and not on the actual food intakes because these have been solved out in the maximization process. While the notion of a health production function is reappraised in section 3.2, it should be noted that few researchers outside economics invoke assumptions such as the optimization behavior of illiterate parents in developing countries for "optimally" selecting health outcomes for their children. On the contrary, nutritionists and psychologists argue that even in an advanced country such as the US, nutrition education programs are necessary to educate individuals about healthy diet and lifestyles (see Chapter 7). Such issues are especially relevant in view of the obesity epidemic in developed countries (Chapter 7). As mentioned in Chapters 1 and 2, economists' definition of rational behavior in making food choices needs to be reconciled with "procedural" definitions used in the psychological literature.

Health functions versus health production functions

Stigler (1945) referred to the relationships between "health inputs" such as nutrient intakes and health outcomes as "health functions". This is a better terminology than "health production functions" since, as discussed below, there are critical differences in the production of the commodities and processes underlying health outcomes. From biomedical and nutritional standpoints, it is important to understand the actual processes affecting health. Such issues have also been recognized in the context of production functions for commodities. For example, Chenery (1949) modeled the process of pipeline transportation by specifying the technical details of the engineering process. Similarly, Johansson (1972) emphasized that inputs are used by firms in "fixed" proportions; once an industrial plant is built, there can be some substitution between capital and "skilled" and "unskilled" labor, though it is constrained in the short run. The critical difference between the Chenery–Johansson approach and that used in the recent economics literature is that while the former incorporated technical information to enrich the model, the latter invokes simplifying assumptions such as the existence of a production function that is twice differentiable with respect to inputs. The use of "health production functions" typically leads to models that explain outcomes such as children's height and weight by food prices and household incomes (e.g. Behrman and Deolalikar, 1988). As noted in Chapter 2, food prices may show little variation when the surveys are confined to narrow geographical regions. In contrast, nutritionists and biomedical scientists are interested in quantifying the effects of specific nutrients on health indicators, partly to corroborate biological hypotheses. The approaches in the biological sciences and economics can be reconciled by invoking a less strong definition of rationality and by modeling individual-level data on nutrient intakes and health outcomes.

Further, there are critical differences between the production of commodities and the manner in which health indicators evolve over time; the time dimensions of the relationships are important for distinguishing between the processes (Bhargava, 1994). In the production of commodities, one can disaggregate the production process over time into the production of primary and intermediate goods that are subsequently used in producing the final output. The "input–output" system of Leontief (1950) for a country, for example, describes the interdependence between various sectors. By contrast, health indicators such as height, weight and sicknesses cannot be viewed as primary, secondary or final "outputs" of a conventional production process. This is because at any point in time, one can simultaneously measure children's heights, weights and sicknesses. However, it is still possible to view children's heights, weights and sicknesses as "stages in health", though the analogy between the production of commodities and health indicators needs to be drawn using a time-dependent framework.

For elaborating on different "stages of health", three indicators are considered, namely, children's height, weight and morbidity levels. Unlike the production of commodities, these stages cannot be simply disaggregated over time. However, the lags underlying health indicators have a natural time ordering. First, the height of a child reflects a history of nutritional intakes and sicknesses; height approximates skeletal size and cannot be altered in the short observation period (e.g. 24 hours) during which the measurements are taken. Second, the weight of a child can vary during the 24-hour period depending on food intakes and mechanisms such as water retention by the body. Moreover, one can predict a child's weight given height (Ehrenberg, 1968), and using the proportions of muscle mass and fats which have different densities (Bhargava, 1999). Third, children's sicknesses, measured (say) in the previous two weeks, can vary dramatically in the observation period due to various symptoms incorporated in morbidity indices. Moreover, resistance to disease is influenced by children's nutritional status, which is reflected in height and weight. Thus, one can view height, weight and morbidity as stages in child health, where these indicators have been arranged using a time-dependent criterion, i.e. they are less "fixed" as one proceeds in this sequence. Alternatively, one can draw analogies for these three health indicators using the economic formulations for "stock" and "flow" variables; height and weight reflect the "stock" of health, while morbidity is a "flow" variable. Such formulations are consistent with the anthropometric and biomedical literatures (Tanner, 1986), and "feedback effects" such as the morbidity of children's lowering body weights can be investigated in this framework.

3.2 The formulation of empirical models for child health

Many aspects of the three health indicators (height, weight and morbidity) discussed above are reflected for Filipino children in the empirical model illustrated in Figure 3.1 (Bhargava, 1994). The formulation of these relationships was influenced by the socio-economic and health variables available in the data set. The salient features of the model were, first, that socio-economic variables such as maternal time allocation and household incomes were postulated to affect intakes of energy and nutrients. Moreover, specific nutrients affect health indicators. For example, energy and protein intakes affect children's weight, while protein, calcium and iron intakes affect height. Similarly, β-carotene and ascorbic acid intakes affect morbidity since these nutrients enhance the immune system. Because children were in the age group 1–10 years, their ages affect dietary intakes. Second, the models for height, weight and morbidity were "triangular" in the sense that height was explained by socio-economic and nutritional factors. Height, in turn, was an explanatory variable in the

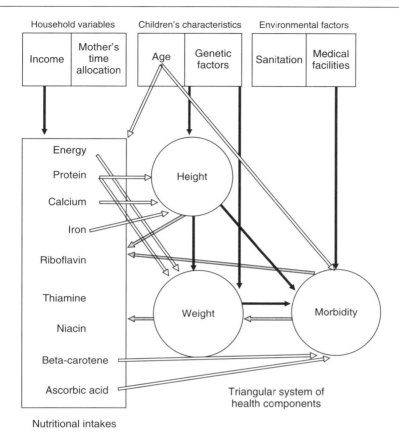

Fig. 3.1 Graphical representation of the empirical model for the health indicators of Filipino children

Source: Bhargava (1994).

model for weight, and height and weight were "stock" variables affecting the "flow" variable, morbidity. The model also recognized that there were likely to be possible feedbacks from health indicators into dietary intakes. For example, current morbidity can lower a child's weight and reduce energy and nutrient intakes. Moreover, morbidity in the current time period can affect subsequent linear growth, though this is not illustrated in Fig. 3.1 to avoid further complicating the presentation. A study investigating the effects of past episodes of diarrhea on children's linear growth (Checkley *et al.*, 2003) is discussed in section 3.4.

Third, children's morbidity levels were affected by their age and environmental variables, such as sanitation, and by access to medical facilities.

Unobserved genetic factors may play a role in determining children's height and weight and can be captured by including anthropometric measures of parents in these models. By contrast, morbidity may be less affected by genetic factors; transmission of bacterial diseases due to an unhygienic environment can overshadow the potential benefits from genetic polymorphisms that may increase children's resistance to disease. Furthermore, several genes are likely to contribute to anthropometric indicators such as height and weight. The main purpose of discussing the possible role of unobserved genetic factors is that one is likely to estimate high between-children variations in models for height and weight due to genetic differences. In contrast, unobserved differences in children's morbidity may be small, especially if the model includes variables reflecting environmental factors such as sanitation. Also, children are likely to develop resistance to disease with age, so that the unobserved between-children differences underlying morbidity may be small in magnitude.

Last, the empirical model in Fig. 3.1 recognized the importance of possible "feedback" effects or endogeneity of variables, such as energy and nutrient intakes and anthropometric indicators, in the relationships. For example, as noted in section 2.2, an individual's energy requirement depends on the basal metabolic rate (BMR) and physical activity levels; BMR in turn is affected by muscle and fat mass (body weight) and by height (James and Schofield, 1990; Johnstone *et al.* 2005). Thus, one would want to include children's height and weight in empirical models for nutrient intakes in order to approximate the energy requirements. In a long time frame, however, nutrient intakes also affect height and weight, so that it may be necessary to treat the intakes as endogenous variables in the models for anthropometric measures. Such issues are seldom discussed in the biomedical literature though they are emphasized in the economics literature on health. However, instrumental variables-type estimators used in economics invoke assumptions that may not be met in practice. The extent to which endogeneity issues can be addressed in empirical models for health indicators is discussed in the next subsection.

Some methodological issues in modeling the relationships between dietary intakes and anthropometric measures

Researchers in the biomedical field typically estimate simple correlations or associations between dietary intakes and health indicators such as height, weight and cognitive test scores (Neumann *et al.*, 1992). There are biological reasons to expect that children in developing countries consuming nutritious foods, such as those high in protein and iron, will attain higher cognitive test scores. In a similar spirit, economists estimate models for health indicators reflecting an underlying "causal" relationship, taking into account several factors. While establishing "causality" is a complex issue (see e.g. Cox, 1992; Adams *et al.*, 2003), the Keynesian dictum in economics that "everything depends on everything else" applies to nutritional sciences because the causal direction in the relationships can be ambiguous. For example, at a given point in time (e.g. on the day of measurement), a child's energy intakes are likely to reflect energy requirements (and expenditures) that depend on anthropometric measures and on physical activity. However, current body weight is also a reflection of past energy and nutrient intakes. Furthermore, if we observe children in areas with endemic food shortages, then children's energy intakes are likely to be below their requirements, i.e. weight may not be a good predictor of energy intakes (Bhargava and Reeds, 1995). Thus, it is reasonable to explain children's energy intakes by their body weights, and also to investigate the effects of nutrient and energy intakes on body weights, i.e. both directions of the relationships are of interest depending on the issues being investigated and the type of data available. Such relationships are also of interest in developed countries in view of the obesity epidemic, where one might define children's energy requirements using their "optimal" rather than actual weights (see Chapter 7).

Further, an "econometric" approach to modeling the effects of energy and nutrient intakes on body weight might treat the intakes as endogenous variables in the estimation. This is partly because food intakes depend on body weight and also because unobserved child characteristics such as fondness for food consumption may increase both the energy intakes and weights. Such factors, in turn, will introduce dependence between errors, affecting the model for weight and energy intakes. One can estimate models for body weights using longitudinal data provided that there are explanatory variables in the model that are not correlated with the errors. Moreover, if only children-specific random effects are assumed to be correlated with energy intakes, then deviations of energy intakes from their time means can also be used as instrumental variables. Alternatively, one can estimate the "reduced form" of the model, which depends on exogenous (or predetermined) variables in the model. However, both these solutions present difficulties in the analysis of nutrition and health relationships. First, unlike aggregate time-series data, correlations between variables in cross-sectional data are often low and it is difficult to postulate joint dependence between uncorrelated variables. Second, in a related vein, exogenous variables used as instrumental variables in the estimation are often poorly correlated with endogenous variables and can lead to worse results in some cases than when the endogeneity is ignored (Bhargava and Sargan, 1983; Bhargava, 1991a). Last, "macroeconomic" variables such as food prices have been used as instruments in previous research using cross-sectional data, though such variables do not aid identification in a longitudinal setting because they are constant or very similar for most households.

The difficulties in modeling relationships between dietary intakes and anthropometric measures suggest that one may need to work with special forms of endogeneity such as postulating that only the random effects δ_i (equation (2.3)) are correlated with the endogenous explanatory variables. Moreover, the null hypothesis that the random effects are uncorrelated with explanatory variables can be tested using likelihood ratio and other chi-square tests that are valid provided the number of individuals in the sample is large. These methods can be illustrated using the data on Filipino children. For example, the model for children's energy intakes estimated in Bhargava (1994: table 1) treated body weight (and height and household's total expenditure) as variables correlated with children-specific random effects. The test statistic was 1.22 and was distributed as a chi-square variable with 12 degrees of freedom. Thus, the exogeneity null hypothesis was accepted by the Filipino data, i.e. it was reasonable to treat body weight as uncorrelated with children-specific random effects. Furthermore, Bhargava (1994: table 3) included children's intakes of energy and protein in the model for body weight but their coefficients were not significantly different from zero (see table 3.2 below). This was also the case when energy and protein intakes were treated as endogenous variables. While the random-effects correlation pattern for exogeneity may seem restrictive, there is likely to be a fair amount of variation in intakes data, making it difficult to explain variables such as height and weight that are measured more accurately. Even so, anthropometric measures such as weight are likely to be important predictors of intakes, and tests for exogeneity are useful for detecting possible misspecifications in the empirical models.

3.3 Empirical models for children's heights and weights

There have been only a few studies estimating "structural" models for children's heights and weights explained by energy and nutrient intakes and other socio-economic and demographic variables. For example, the Cebu Study Team (1992) estimated a model for the weights of young Filipino children, controlling for patterns of breast-feeding and weaning. Also, indicators for children's morbidity due to diarrhea and febrile respiratory infections were included as explanatory variables in the models. However, energy and nutrient intakes and children's heights were not included in the model. As noted in section 3.2, height reflects skeletal size and is widely used in the anthropometric assessment literature to predict weight (Cole, 1991). In fact, one can include height as a potentially endogenous variable that is correlated with the children-specific random effects (δ_i in equation (2.3)) in the model. Empirical results for children's heights and weights using data from the Philippines and Kenya are presented in the next subsection; certain empirical models for children's morbidity are discussed in section 3.4.

Empirical results from models for Filipino and Kenyan children's heights

Empirical models of the type described in Fig. 3.1 for children's heights and weights were estimated by Bhargava (1994) using longitudinal data on 312 Filipino children aged 1–10 years observed in four time periods at four-monthly intervals. Similar models were estimated for 102 Kenyan school children who were observed in three time periods. The results for Filipino children's heights are presented in Table 3.1; findings from the model for Kenyan children's heights are also summarized. The results for Filipino children's heights are presented for two specifications, namely, where the intakes of protein and calcium were treated as endogenous variables and where they were assumed to be exogenous. Because the model is dynamic, containing the previous or lagged height, the form of endogeneity for nutrient intakes assumes that they are correlated only with the random effects. The main findings were, first, that the results from the models assuming nutrient intakes to be endogenous and exogenous were similar. While the exogeneity hypothesis was rejected at the 5% significance level, it was accepted at the 2.5% level.

Second, children's age was estimated with a positive coefficient that was significant in both specifications. Indicator variables for survey rounds 3 and 4 were also significant. Moreover, parental heights were significant and the two coefficients were close. However, the null hypothesis that the coefficients of maternal and paternal heights were the same was rejected by the data; the chi-square statistic, distributed with one degree of freedom, was 7.11 and was higher than the 5% critical limit of 3.84.

Table 3.1 Maximum-likelihood estimates for heights of Filipino children

Parameter	Nutrient intakes endogenous[a]	Nutrient intakes exogenous
Constant	0.126	0.120
	(0.122)	(0.096)
Age	0.030*	0.020*
	(0.004)	(0.006)
Father's height	0.063*	0.050*
	(0.018)	(0.011)
Mother's height	0.067*	0.044*
	(0.019)	(0.012)
Mother's housework	0.0003	0.002*
	(0.0005)	(0.0006)
Protein	0.001	0.002*
	(0.001)	(0.001)
Calcium	−0.0002	0.0013
	(0.0010)	(0.0010)
Time period 3	0.005*	0.004
	(0.001)	(0.003)
Time period 4	0.0003*	0.003*
	(0.0010)	(0.001)
Lagged dependent variable	0.813*	0.805*
	(0.017)	(0.024)
Chi-square (8)[b]	17.67*	
Chi-square (8)[c]		62.94*
Chi-square (1)[d]		7.11*

Note: 312 children were observed, in 4 time periods; figures in parentheses are standard errors.
[a] The intakes of protein and calcium were assumed to be correlated with random effects.
[b] Likelihood ratio statistic for exogeneity of protein and calcium intakes.
[c] Likelihood ratio statistic for the random-effects decomposition.
[d] Likelihood ratio statistic for the null hypothesis that coefficients of parents' heights were the same.
* $P < 0.05$.
Source: Bhargava (1994).

Third, mothers' housework was a significant predictor of children's heights. Moreover, children's protein intakes were positively and significantly associated with heights in the specification that assumed intakes to be exogenous. A few methodological issues should be mentioned in this context. Because 24-hour recall data exhibit large within-subject (intra-individual) variation, intakes in each survey round may not reflect the "habitual" intakes. An alternative would be to average the nutrient intakes over the four survey rounds for which the data were available and introduce average intakes as time-invariant explanatory variables in the model. Using this procedure, both protein and calcium intakes were significantly associated with children's heights. This issue will be further discussed using the data from Kenya, where nutrient intakes in each survey round were based on six days of food intakes. It is important to have accurate information on children's intakes. Issues of endogeneity of intakes are also important; however, changes in the estimated model parameters when endogeneity was addressed were small in the model for Filipino children's heights.

Fourth, coefficients of the lagged dependent variables were large ($\sim$0.80) and highly significant, indicating that the long-run effects of explanatory variables were six times that of the respective short-run impacts in Table 3.1. This was not surprising since height increases gradually over time and hence the "equilibrium" effects of nutrient intakes are likely to be greater. Lastly, the simple random-effects decomposition for error terms (as in equation (2.3)) was rejected by the data; the chi-square statistic with 8 degrees of freedom was 62.94. This was perhaps not surprising since the transitory components (v_{it}) of the errors were likely to be serially correlated and their variances can change over time. Thus, it was useful to invoke the assumption that u_{it} were distributed as a multivariate normal distribution; restricting the errors to have the simple random-effects decomposition would lead to inconsistent estimates of model parameters.

The results for children's heights from Kenya were similar to those for Filipino children's heights in that the coefficient of the lagged dependent variable was 0.91, i.e. children's heights evolved very gradually over time and the long-run effects were ten times as high as the short-run impacts (Bhargava, 1999). While children's intakes of calcium were positively associated with heights, the coefficient was significant at only the 10% level. Thus, in spite of the considerably more accurate information compiled in the Kenyan data on children's nutrient intakes, calcium intakes did not predict children's heights. Of course, the coefficient might have been significant if the data were available on a greater number of children or if children were observed in more than three time periods. It is evident that the study design is important for extracting information from the data, though variations in dietary intakes are often increased in developing countries because of poverty (Bhargava, 1992). In fact, the famine in Kenya in 1984 during the study period (Cohen and Lewis, 1987) exacerbated the variations in food intakes.

Last, maternal heights and household incomes in Kenya were significantly positively associated with children's heights, whereas a greater number of children in the household was significantly negatively associated. The quality of diet in households depends on economic factors and is particularly likely to deteriorates if several young children are born to the mother within a short time interval. From a policy standpoint, it is important to improve children's diet quality to enhance their growth. Thus, it would be useful to collect more accurate dietary information by recording several days of intakes, with sample sizes sufficiently large to afford precise estimation of the magnitudes of the effects. Notwithstanding the methodological difficulties, the results from the Philippines and Kenya provided some evidence concerning the benefits of higher intakes of protein and calcium for children's linear growth.

Table 3.2 Maximum-likelihood estimates for weights of Filipino children

Parameter	Results with parents' height and weight[a]	Results with BMI or weight[b]
Constant	−0.908*	−0.832*
	(0.149)	(0.142)
Age	0.001	0.001
	(0.008)	(0.009)
Father's height	−0.116*	
	(0.011)	
Mother's height	0.024	
	(0.017)	
Father's weight	0.055*	
	(0.016)	
Mother's weight	0.025*	0.027*
	(0.013)	(0.013)
Father's BMI		0.055
		(0.031)
Mother's housework	0.003*	0.003*
	(0.001)	(0.001)
Height	0.373*	0.374*
	(0.026)	(0.088)
Current morbidity	−0.006*	−0.006*
	(0.002)	(0.002)
Energy	0.007	0.007
	(0.007)	(0.007)
Protein	−0.002	−0.002
	(0.006)	(0.007)
Time period 3	0.015*	0.015*
	(0.006)	(0.006)
Time period 4	−0.021*	−0.002
	(0.005)	(0.006)
Lagged dependent variable	0.736*	0.736*
	(0.027)	(0.055)
Chi-square (2)[c]	0.52	
Chi-square (8)[d]	56.30*	56.71*

Note: 312 children were observed, in 4 time periods; figures in parentheses are standard errors.

[a] The coefficients of parents' heights and weights were assumed to be unrestricted.

[b] Father's height and weight combined as the BMI (weight/height^2) and mother's height was dropped.

[c] Likelihood ratio statistic for the null hypothesis that father's height and weight can be combined as BMI and mother's height can be dropped.

[d] Likelihood ratio statistic for random-effects decomposition.

* $P < 0.05$.

Source: Bhargava (1994).

Empirical results from models for Filipino and Kenyan children's weights

The results from the model for Filipino children's weights are presented in Table 3.2 for two specifications (Bhargava, 1994). In the first specification, parents' heights and weights appear as separate explanatory variables. In the second specification, paternal height and weight are combined as the BMI, and maternal height is dropped and maternal weight is retained as an explanatory variable. The statistical test for combing height and weight as the BMI is discussed in section 3.4 in the context of children's morbidity. The main findings in Table 3.2 were that there were significant positive associations between children's weight and parental heights and weights. The model controlled for children's ages; the results were not surprising since food intake patterns for household members as well as genetic factors were likely to lead to cross-generational associations in anthropometric measures. Second, maternal housework was a significant predictor of body weight. It should be noted that the housework variable was constructed by inquiring into the effort mothers put in to housework in the previous four-month period. Thus, while it is possible that mothers occasionally increased their housework in response to greater needs such as those arising from children's sicknesses, it is appropriate to treat housework as an exogenous variable.

Third, children's height was an important predictor of weight, and the estimated coefficient of 0.37 was close to the coefficients estimated by Ehrenberg (1968) using simple models that regress the logarithm of weight on height. These similarities reflect the underlying relationship between height, skeletal size and skeletal weight. Fourth, children's morbidity was estimated with a negative coefficient that was significant at the 5% level; children that had been sick in the previous fortnight had lower body weight. However, energy and protein intakes were not significant predictors of weight. While this may seem surprising, the large variation in 24-hour recall data may be partly responsible for these findings. Moreover, quality of diet is often reflected in the ratio of protein to energy intakes and this issue is discussed using the Kenyan data. Last, the estimated coefficient of the lagged dependent variable was 0.74, which was quite large, though smaller than the corresponding coefficient in the model for height. Thus, the long-run impacts of explanatory variables on weight were about 1.33 times the short-run coefficients. The simple random-effects decomposition for the errors was rejected in the model for weights.

The main insight afforded by the empirical model for Kenyan children's weights (Bhargava, 1999) were that, first, children's arm circumference was a significant predictor and estimated with a large coefficient (0.60). The estimated coefficient of height in the model was 1.07, which was considerably larger than the estimate from the model for Filipino children's weights. Arm circumference reflects muscle mass, which has a higher density than body fat and is a useful measure to include in models for weight. Second, when protein and energy intakes were included as separate regressors in the model for children's weights, the sign of protein intakes was positive, while that of energy intakes was *negative*. A likelihood ratio test accepted the restrictions (see section 3.4) that protein and energy intakes should be expressed as the ratio of protein to energy intakes. As noted in Chapter 2, this ratio is a good indicator of diet quality; Kenyan children consuming higher quantities of protein relative to energy intakes were significantly heavier. This is perhaps not surprising, since poor households are likely to consume large quantities of staple foods to meet their energy needs, whereas protein intakes are likely to be higher among better-off households. Even so, it is remarkable that econometric models for children's body weights corroborated the knowledge in the nutritional sciences regarding the importance of the protein–energy ratio for children's physical growth. Overall, Filipino children were better nourished than Kenyan children, though the qualitative results from the two sets of analyses were quite similar.

3.4 Empirical models for children's morbidity

While children in developing countries are frequently sick with diarrhea, respiratory infections, and other ailments, it is common for social scientists such as demographers to analyze the proximate determinants of child mortality. Of course, mortality is the extreme form of illness and is preceded by sicknesses. Moreover, connections between nutritional status and infection are recognized by biomedical scientists (see Scrimshaw *et al.*, 1959) and children's morbidity often affects their physical growth and immunological functions. The difficulties in compiling data on child morbidity include the fact that the number of children in the sample is limited by the resources available for the studies and the fact that ethical considerations require that seriously ill children receive appropriate treatments. By contrast, demographic surveys compile *ex post* information in developing countries on child mortality for large numbers of households, thereby obviating the need for treatment against diseases. In Chapter 5, the proximate determinants of infant mortality will be investigated using a demographic survey from India. However, it is also important to analyze proximate determinants of child morbidity, since it is imperative that undernourished children receive appropriate interventions to enhance their resilience to disease, which, in turn, affects physical and intellectual development. While it is difficult to develop general frameworks for analyses of morbidity data, the results for the morbidity indices of Filipino and Kenyan children (which were included in the analyses of height and weight data in section 3.3) provide useful insights. In addition, analyses of child morbidity in Bangladesh, Pakistan and Peru are briefly discussed to provide the reader with a broader view of these issues.

Empirical results from morbidity indices for Filipino and Kenyan children

As noted in section 3.1, morbidity indices constructed from the number of symptoms and days for which a child is sick provide better measures of intensity and duration of sicknesses than indicator (0/1) variables. Bhargava (1994) developed a model for Filipino children's morbidity index, based on the number of days on which the child was sick in the last two weeks with five symptoms (cold, cough, fever, diarrhea and headaches):

$$
\begin{aligned}
\ln(\text{Morbidity})_{it} = {} & b_0 + b_1 \ln(\text{Age})_i + b_2 \, (\text{Open pit toilet})_i \\
& + b_3 \, (\text{Distance from facility})_i + b_4 \, (\text{Mother's housework})_{it} \\
& + b_5 \ln(\text{Height})_{it} + b_6 \ln(\text{Weight})_{it} \\
& + b_7 \ln(\beta\text{-carotene intake})_{it} + b_8 \, (\text{Time period3})_t \qquad (3.2) \\
& + b_9 \, (\text{Time period4})_t + b_{10} \, (\text{Morbidity})_{it\text{-}1} \\
& + u_{it} \quad (i = 1, \ldots, 312; \, t = 1, \ldots, 4)
\end{aligned}
$$

In this model, children's height and weight appear as separate regressors and one can test whether these variables should be combined as the BMI if a likelihood ratio test cannot reject the null hypothesis:

$$
b_5 + 2b_6 = 0 \qquad (3.3)
$$

If the null hypothesis is rejected, then it would be better to include children's heights and weights as separate regressors. Similarly, one can test whether height and weight can be combined as weight-by-height if the coefficients b_5 and b_6 are equal with opposite signs. The advantage in testing these restrictions is that if the null hypothesis is rejected, then the estimated parameters from models containing the BMI (or

Table 3.3 Maximum-likelihood estimates for morbidity of Filipino children

Parameter	Results with height and weight[a]	Results with BMI[b]
Constant	−16.564*	− 15.642*
	(3.880)	(3.169)
Age	−0.528*	−0.462*
	(0.172)	(0.046)
Open pit toilet	0.177*	0.174*
	(0.089)	(0.088)
Distance[c]	0.059*	0.060*
	(0.017)	(0.017)
Mother's housework	0.054	0.055
	(0.037)	(0.036)
Height	4.374*	
	(1.142)	
Weight	−2.020*	
	(0.474)	
BMI		−2.069*
		(0.495)
β-carotene	−0.035*	−0.034*
	(0.019)	(0.017)
Time period 3	−0.249*	−0.239*
	(0.100)	(0.099)
Time period 4	−0.103	−0.089
	(0.102)	(0.093)
Lagged dependent variable	0.051	0.045
	(0.046)	(0.047)
Between/within variance	0.061	0.067
	(0.043)	(0.044)
Within variance	1.519	1.517
Chi-square (l)[d]	4.86	
Chi-square (8)[e]	17.32*	16.47*

Note: 312 children were observed, in 4 time periods; figures in parentheses are standard errors.
[a] The coefficients of children's height and weight were assumed to be unrestricted.
[b] Height and weight combined as BMI (weight/height^2).
[c] Average time of travel (in minutes) to nearest hospital, paramedic facility and doctor.
[d] Likelihood ratio statistic for combining height and weight as BMI.
[e] Likelihood ratio statistic for random-effects decomposition.
* $P < 0.05$.
Source: Bhargava (1994).

weight-by-height) as a regressor are no longer consistent. This is especially important in situations where the coefficients of height and weight have the same signs and hence it is altogether inappropriate to combine them as BMI (or weight-by-height).

The results for Filipino children's morbidity are presented in Table 3.3 for two specifications, namely, where children's height and weight were introduced as separate regressors and where these variables were combined as the BMI. The salient findings in Table 3.3 were that the coefficient of height and weight were opposite in sign, and the likelihood ratio tests accepted the null hypothesis that height and weight can be combined as the BMI. In fact, the coefficient of height was almost exactly twice as large in absolute terms as that of weight. Thus, in modeling the morbidity levels of Filipino children, BMI was a good measure of health status and is likely to identify children that are susceptible to sicknesses. Note, however, that the underlying causality can also run in the reverse direction, i.e. sicker children had lower BMI. This issue is further discussed using the Kenyan data, which allow greater flexibility in modeling due to the more elaborate morbidity questionnaires.

Second, children's age was estimated with a large negative coefficient, indicating that older children were sick less frequently, presumably due to the immunity acquired to infections with age. Distance of the household from the nearest medical facility was estimated, with a positive and significant coefficient indicating the importance of access to medical care for achieving less sickness. Third, maternal housework was not a significant predictor of children's morbidity. However, the intakes of β-carotene were significantly negatively associated with children's morbidity. This last was an important finding, especially since diet quality in developing countries is likely to suffer due to fluctuations in prices of fresh fruits and vegetables.

Fourth, the lagged dependent variable was not significantly different from zero; this was perhaps not surprising, since patterns in child morbidity were less likely to exhibit time dependence. Last, the chi-square statistic for testing the random-effects decomposition was close to its critical level. Thus, the between-and within-subject variances were also estimated for the model for children's morbidity. However, the between–within variance ratio was not statistically significant. There were insignificant unobserved between-children differences in morbidity and this may be due to the fact that the children's anthropometric indicators, nutrient intakes and environmental variables explained much of the variation in morbidity.

The morbidity data on Kenyan children were considerably more detailed and covered the previous three-month period for recording the duration of various sicknesses. In addition, blood indicators of iron status, such as children's hemoglobin concentration, were measured, and the dietary intakes at each of the time points were based on six days of food intakes. The results for the morbidity index of Kenyan children are reported in Table 3.4. First, children's hemoglobin concentration was significantly negatively associated with morbidity levels. Hemoglobin concentration is a good measure for iron status and, unlike other measures such as serum ferritin, it is not elevated in children with inflammations or

Table 3.4 Maximum-likelihood estimates of the model for the morbidity index of Kenyan children

Variable/Parameter	Estimated coefficients
Constant	21.554*
	(122.47)
Hemoglobin	−0.594*
	(16.97)
Maternal age	−0.924*
	(16.21)
Paternal score on cognitive tests	−0.219*
	(4.87)
Maternal score on cognitive tests	−0.835*
	(23.19)
No latrine	1.599
	(1.70)
SES+ Cash Income	−3.284*
	(15.42)
(SES+ Cash Income)[a]	2.415*
	(14.56)
Vitamin A	−0.240*
	(8.57)
Maternal morbidity	0.238*
	(4.33)
BMI	−2.673*
	(46.09)
Lagged dependent variable	0.075
	(0.95)
Between/within variance[a]	0.151
	(1.10)
Within variance	1.395
Chi-square(9)[b]	5.22
Sample size	102

Note: 103 children were observed, in 3 time periods; absolute values of asymptotic t-statistics are in parentheses.
[a] Random-effects specification was accepted.
[b] Chi-square test for exogeneity of vitamin A intakes, maternal morbidity, and BMI.
* $P < 0.05$.
Source: Bhargava (1999).

infections (Bhargava, Jukes *et al.*, 2003a). Good diet quality in terms of the intake of bioavailable iron is essential for raising hemoglobin concentration. Second, the paternal and maternal scores on cognitive tests were negatively and significantly associated with children's morbidity. The coefficient of maternal test scores was four times as high as that of paternal scores, which was not surprising since women are mostly involved in caring for children. Moreover, years of maternal and paternal education were not significant predictors of children's morbidity, in part because most adults had never attended school.

Third, vitamin A intakes were negatively associated with children's morbidity. This finding complements the results for the Filipino children, where β-carotene intakes were associated with lower morbidity (β-carotene is converted to vitamin A by the human body). Fourth, children's BMI was negatively associated with their morbidity levels. Moreover, replacing the current BMI with lagged BMI led to very similar results. Alternative specifications were estimated separating the height and weight variables though the results were similar. Finally, the lagged dependent variable and between–within variance ratio were not significant predictors of Kenyan children's morbidity levels. These results were in agreement with those for Filipino children in Table 3.3 and provide evidence that socio-economic variables and measures of nutritional status explain a substantial amount of variation in children's morbidity levels in developing countries.

Some related models for child morbidity

Gastrointestinal diseases such as diarrhea, dysentery, cholera and typhoid are common in developing countries because of poor sanitation and frequent exposure to pathogens (Keusch *et al.*, 2006). Child morbidity is particularly likely to increase as young children are weaned. Simple interventions such as washing hands with soap and storing water in containers with narrow mouths can reduce the concentration of pathogens. For example, in a randomized trial in Karachi, Pakistan, 25 neighborhoods were assigned to a program promoting hand washing, while 11 neighborhoods served as controls (Luby *et al.*, 2005). The study design created three groups of households, i.e. those supplied with anti-bacterial soap, those given plain soap, and controls, with approximately 300 households in each group. The morbidity outcomes included the incidence of pneumonia, impetigo and diarrhea. The main findings were that children under 5 years in households that received plain soap had a 50% lower incidence of pneumonia than the control group. Moreover, children under 15 years had a 53% lower incidence of diarrhea, and 34% less impetigo compared to the control group. All these differences were statistically significant at the 5% level. However, disease incidence did not differ significantly between the households that received anti-bacterial soap and those that received plain soap. It is evident that simple interventions such as washing hands with soap can have dramatic effects for lowering bacterial diseases in developing countries. Moreover, prevention of gastrointestinal diseases reduces nutrient loss and economizes on resources required for improving children's health and nutritional status via dietary intakes.

Further, episodes of diarrhea and other sicknesses can lower children's physical growth. For example, a study by Checkley *et al.* (2003) found that among 225 Peruvian children followed from birth to 35 months, children that had diarrhea 10% of the time during the first 24 months were 1.5 cm shorter than their counterparts without diarrhea. Moreover, the adverse effects of diarrhea differed by age; diarrhea in the first six months appeared to result in long-term height deficits. However, food intakes were not measured in this study; it is plausible that the deficits in heights need not be permanent and early medical interventions and targeted food supplementation programs can help to reduce growth deficits. At any rate, lowering the prevalence of diarrhea should be a goal of health policies through improvements in sanitation, though such interventions are often expensive and beyond the budgets of local governments.

Finally, there have been studies investigating the effects on children's morbidity of contamination of water by coliforms such as fecal coliforms and *E. coli*. For example, van Derslice and Briscoe (1995) used data on 2,355 Filipino infants and argued that keeping the water at the source free of contamination was more important than storage methods at home. In contrast, Moe *et al.* (1998) underscored the importance of pathogens in the stored water. Of course, it is important to reduce water contamination at the source as well as in storage. In a study in Bangladesh, Bhargava, Bouis *et al.* (2003) created an index for children's gastrointestinal morbidity and found that the concentration of fecal and total coliforms in the water stored in the home significantly increased children's morbidity. Moreover, coliforms in the water at the source, as well as poor storage methods and hygiene practices, increased colonies of coliforms in the stored water. Thus, in addition to interventions such as regular hand washing, it is important to supply households with better storage containers with tight-fitting lids, and to treat the water at the source with hypochlorite solution to reduce contamination. While such interventions may seem expensive, prevention of morbidity not only enhances children's nutritional and health status, but is also important for reducing parental time devoted to children's sicknesses and convalescence.

3.5 Conclusions

In this chapter, the role of socio-economic, nutritional and environmental factors on indicators of child health such as height, weight and sicknesses was discussed. The approaches used in the biomedical field were contrasted with those in economics and an integrated framework was proposed for estimating comprehensive empirical models for health indicators. It was seen that econometric techniques can be applied for tackling the endogeneity of explanatory variables in the health relationships. Moreover, by invoking assumptions that are consistent with the biomedical literature, the econometric approach can incorporate the framework in the anthropometric assessment literature. While food prices affect consumption, the effects of food prices are likely to be reflected in dietary intakes. For example, policy-makers are interested in knowing the effects of intakes of protein, calcium and iron on child growth. Thus, models for children's health indicators that take into account the relationships between the indicators and dietary intakes are useful for food policy formulation.

The importance of diet quality was evident in the models for Filipino children, where protein and calcium intakes were significant predictors of height, and β-carotene intakes were associated with lower morbidity. Similarly, the protein–energy ratio was a significant predictor of weights of Kenyan children, and vitamin A intakes were associated with lower morbidity. The empirical models also highlighted the role of environmental factors such as sanitation for children's morbidity. The randomized trial conducted in Pakistan by Checkley *et al.* (2003) corroborated the importance for reducing morbidity of simple strategies such as washing hands. Moreover, the presence of coliforms in the water in Bangladesh was seen to increase child morbidity. It is important that food policies take a broad view of child health in developing countries, and that children achieve their full physical and cognitive potential. The analytical framework developed in this chapter underscores the connections between nutrient intakes, environmental variables and health outcomes. It is important to assess jointly the costs of various interventions, and formulation of policies would benefit from further empirical analyses for countries. However, poor sanitation is likely to remain a formidable challenge in developing countries because of the high costs of sewage treatment, though policy-makers need to take a long-term view of the health benefits of improving sanitation. Moreover, high population density can exacerbate the transmission of diseases; issues of population growth are discussed in Chapter 5.

4

Child health and cognitive development in developing countries

4.1 Background issues

This chapter extends the framework developed in Chapter 3 for the analysis of health indicators (height, weight and morbidity) to children's psychological indicators such as scores on cognitive tests. This is important because children's learning is a cumulative process and future supply of skilled labor depends on cognitive achievements at various stages; it is known that economic development depends on the availability of a skilled labor force, which is critical for production of goods and services (see e.g. Schultz, 1961). Developing countries where a majority of children are under-nourished and without primary school education are unlikely to be competitive in the age of globalization. Moreover, goods and services demanded in developed countries typically entail high added value. While one can improve the productivity of unskilled labor by improving their nutritional intakes, it is not possible to train illiterate individuals to perform tasks requiring scientific skills. In order to enhance the productivity of a developing economy, therefore, policy-makers need to understand the mechanisms underlying children's cognitive development and incorporate the empirical findings in policy formulation (Bhargava, 2001a).

The factors underlying children's physical growth and cognitive development in developing countries are intertwined and have been investigated chiefly by nutritionists and psychologists. The emphasis in these studies has been on issues such as the effects of maternal nutritional and health status on fetal growth, the dynamics of infant and child growth reflected in height and weight, and on the cognitive development of school-aged children. While some studies have been randomized controlled trials driven by specific hypotheses such as the effects of food supplementation on birth outcomes (e.g. Ceesay *et al.*, 1997), the empirical findings can be interpreted in a broad framework for policy formulation.

Further, health and education policies for enhancing child development critically depend on children's ages. For example, if pregnant women are severely under-nourished, then high priority should be given to food supplementation programs for them, in order to prevent intra-uterine growth retardation. By contrast, if children are mildly under-nourished and can benefit from instruction in school, then investments in the educational infrastructure may bring high returns. While such issues are recognized by researchers, biomedical investigators, in view of the complexities of the problems, are often forced to focus on very specific hypotheses. For example, nutritionists designed a longitudinal study in Guatemala spanning several years to study the effects of a supplement high in protein on child growth and cognitive development (Pollitt *et al.*, 1993; Martorell and Scrimshaw, 1995). The study design was guided by the emphasis on protein intakes for brain development (Monckeberg, 1975). However, the recent nutrition literature has emphasized the role for cognition of micronutrients such as iron. In fact, the control group in the Guatemala study received an iron-fortified drink ("Fresco") that might have inadvertently narrowed the differences between the two groups in children's scores on cognitive tests. Because knowledge is constantly evolving in the biomedical sciences, it is important for social scientists to be familiar with the scientific literature. Biomedical researchers, in turn, should develop comprehensive study designs so that the data emerging from their studies can be analyzed for policy formulation.

Because randomized trials are becoming popular in the social sciences, it is important to develop a broad framework for the analysis and interpretation of data from control and treatment groups. As noted in section 4.4, population-based interventions typically affect a range of variables and it is important to understand the various linkages and pathways. Section 4.2 discusses certain studies investigating the effects of maternal health status on infant development. The growth of pre-school children is discussed in section 4.3. sections 4.4 and 4.5 focus on the cognitive development of school-aged children and incorporate the developmental psychology literature, while also highlighting the strengths of the data from randomized controlled trials.

4.2 Maternal health status and infant development in developing countries

The importance of maternal nutritional and health status for infant development is widely recognized in the biomedical literature. Kramer (1987) and Bhutta *et al.* (2005) surveyed numerous community-based interventions for promoting infant growth and development. Intra-uterine growth retardation is common in developing countries, where approximately 75% of infants are small for their "gestation age". Improving maternal health status is likely to enhance intra-uterine growth (Falkner *et al.*, 1994; Kallberg *et al.*, 1994). Furthermore, infant length, weight and head circumference at birth are useful indicators that may be differentially affected by maternal health status and environmental factors. For example, children in developing countries, at birth are about the same length as their counterparts in developed countries, though their weight is lower partly because head circumference is smaller. Because the head circumference of an infant is a rough indicator of brain mass (Dobbing and Sands, 1978), poor maternal health status can compromise children's cognitive development in later years. However, the "plasticity" of the human brain to perform various functions (Purves, 1988) can alleviate some of the adverse consequences of maternal under-nutrition. Thus, programs for reducing maternal under-nutrition and for enhancing children's physical growth and learning are essential for children to attain their full cognitive potential.

Maternal nutritional status depends on a variety of socio-economic factors and in turn gradually affects fetal growth. For example, nutritious foods consumed during the first trimester of pregnancy may have greater beneficial effects on fetal growth than in the third trimester, when the rate of infant growth is reduced (Falkner *et al.*, 1994). Moreover, maternal calcium and iron stores are important for fetal growth and are likely to be depleted if a woman gives birth to several children within a short time interval. It is perhaps not surprising that demographic studies in developing countries find short birth intervals to be predictors of higher infant mortality (e.g. Hobcraft *et al.*, 1984; demographic issues affecting child health will be discussed in Chapter 5). Furthermore, randomized controlled trials of food supplementation have shown significant benefits for outcomes such as birth weight and head circumference in Gambia (Ceesay *et al.*, 1997). While small-scale studies of food supplementation for pregnant women will the generally show beneficial effects, development of large-scale interventions entails the incorporation of socio-economic circumstances of households and requires large investments in health care and family planning services.

This chapter begins by presenting results from analyses of additional variables from the survey in the Embu region of Kenya (described in Chapters 2 and 3) on birth outcomes such as length, head circumference and weight (Bhargava, 2000). This longitudinal survey followed women through pregnancy, monitoring their food intakes and compiling socio-economic and demographic information. Moreover, the effects of maternal nutritional status and other relevant factors on the dynamics of infant length, weight and head circumference in the first six months are discussed. Such analyses are seldom performed in the biomedical field, in part because econometric methods are necessary for extracting information from longitudinal data. The two sets of analyses for Kenyan infants summarized in the next subsection provide useful insights into nutritional and socio-economic factors affecting infant development in countries where under-nutrition is prevalent.

Empirical results for Kenyan infants' weight, length and head circumference at birth

The results from estimating models for birth weight, length, and head circumference of 100 infants born in the 1984/5 period in Embu region of Kenya are presented in Table 4.1. The gestation period was estimated using

Table 4.1 Regression models explaining Kenyan infants' birth weight and length and head circumference (measured shortly after birth) by maternal nutritional status and demographic characteristics

| | Dependent variable | | | | | |
| | Birth weight (kg) | | Length (cm) | | Head circumference (cm) | |
	Coefficient	SE	Coefficient	SE	Coefficient	SE
Independent variable						
Constant	−4.280*	0.984	2.620*	0.363	2.487*	0.249
Indicator for sex[a]					0.022*	0.006
Dubowitz (w)	1.068*	0.249	0.285*	0.094	0.168*	0.064
Maternal pre-pregnancy BMI $(kg/m^2)^b$	0.274*	0.104	0.094*	0.038	0.101*	0.026
Maternal hemoglobin $(g/1)^b$	0.136	0.093	−0.013	0.034	0.033	0.023
Maternal hookworms $(g)^b$	−0.014	0.009	−0.007*	0.003	−0.004	0.002
Parity	0.007	0.005	−0.0003	0.0019	0.012*	0.0004
Parity squared					−0.0008*	0.0004
Indicator for birth order 1	−0.102*	0.051	−0.007	0.019		
Infant age (days)			0.002*	0.0004	0.002*	0.0004
Adjusted R^2	0.33		0.27		0.51	
Sample size (n)	102.0		99.0		99.0	

Note: Values are slope coefficients and standard errors.
[a] Boy = l; girl = 0.
[b] These variables and the dependent variable are in logarithms; coefficients of these explanatory variables are elasticities.
*P <0.05.
Source: Bhargava (2000).

the Dubowitz and Dubowitz (1977) method and was transformed into natural logarithms; maternal pre-pregnancy BMI and hemoglobin concentration were also transformed into logarithms. Because not all infants were measured just after birth, the models controlled for age in number of days. The main findings in Table 4.1 were that the elasticity of birth weight with respect to the gestation period was close to unity and was statistically significant; elasticities of infant length and head circumference with respect to gestation period were 0.29 and 0.17, respectively. Thus, a longer gestation period had the highest impact on weight, followed by length and head circumference. Second, the elasticity of birth weight, length and head circumference with respect to maternal pre-pregnancy BMI were all significant, with elasticity for birth weight being the highest (0.27). Thus, good initial maternal nutritional status was a significant predictor of birth outcomes for Kenyan infants.

Third, the indicator variable for whether the infant was a boy was significant in the model for head circumference; gender differences were insignificant for birth weight and length. While maternal hemoglobin concentration was not a significant predictor of the three birth outcomes, maternal hookworm load was negatively and significantly associated with infant length. Anemia is a risk factor for birth outcomes and the effects of maternal hemoglobin concentration on the dynamics of infant growth in the first six months of life will be discussed in the next subsection. Fourth, the birth order of the infant ("parity") was not a significant predictor of birth weight and length. However, the relationship between parity and head circumference was a quadratic one and both the coefficients were statistically significant. The indicator variable for whether the infant was the first-born child was significant only in the model for birth weight. While one is more likely to observe adverse effects of repeated pregnancies on indicators such as head circumference, larger sample sizes would be necessary to investigate the potentially non-linear effects of parity on birth outcomes. Last, food intakes by the mother during the trimesters were not significantly associated with birth outcomes; spells of morbidity during the pregnancy were also not significant. These results may be partly due to the small number of infants in the sample.

Overall, the results in Table 4.1 support the importance of good maternal nutritional status for birth outcomes; additional intakes, especially of animal products, are likely to enhance maternal nutritional status, reflected in BMI. While it is often difficult in rural areas of developing countries to increase food intakes of pregnant women, food supplementation in a timely manner should be a major goal of policies for reducing energy, iron and iodine deficiencies. Early interventions can obviate the need for supplementation programs for children in later years and are likely to be cost-effective because the adverse effects of under-nutrition in early years may be irreversible.

Dynamics of Kenyan infants' length, weight and head circumference in the first six months

While the proximate determinants of birth outcomes are of considerable interest, environmental factors affecting infant growth in the first few

Table 4.2 Maximum-likelihood estimates of dynamic random-effects model for Kenyan infants' length, head circumference and weight in the period from 1 to 6 months, explained by maternal nutritional status and infant nutrient intakes, morbidity, and length and arm circumference

| | Dependent variable | | | | | |
| | Length (cm) | | Head circumference (cm) | | Weight (kg) | |
	Coefficient	SE	Coefficient	SE	Coefficient	SE
Independent variable						
constant	2.466*	0.224	0.479*	0.185	−5.066*	0.545
Sex[a]	0.009*	0.003	0.007*	0.002	0.006	0.008
Maternal hemoglobin (g/1)[b]	0.011	0.013	0.018	0.011	0.077*	0.026
Maternal BMI (kg/m^2)[b]	0.025	0.015	−0.0002	0.009	0.087*	0.019
Paternal BMI (kg/m^2)[b]					0.023	0.025
Maternal head circumference (cm)[b]			0.136*	0.048		
Maternal morbidity index[b]			−0.0006	0.0005	0.0009	0.0026
Infant calcium intake (mg/day)[b]	0.001*	0.0006			0.002	0.002
Infant morbidity index[b]	−0.0001	0.0008	−0.0003	0.0006	−0.006*	0.002
Infant length (cm)[b]			0.211*	0.033	0.973*	0.118
Infant arm circumference (cm)[b]					0.666*	0.060
Indicator time period 3	0.024*	0.005				
Indicator time period 4	0.041*	0.007				
Indicator time period 5	0.061*	0.009				
Indicator time period 6	0.068*	0.011				
Lagged dependent variable (cm)[b]	0.356*	0.059	0.475*	0.039	0.283*	0.039
Between–within variance	0.564*	0.161	0.268*	0.088	0.008	0.047
Within variance	0.0006		0.0003		0.0061	
Chi-square[c]	9.52		18.83		24.14	
2X log-likelihood function	−4375.03		−4371.85		−2469.18	
Sample size (n)	102		92		81	

Note: Values are slope coefficients and standard errors; the infants were observed 6 times at monthly intervals.
[a] Boy = 1; girl = 0.
[b] This variables and dependent variable in logarithms.
[c] Likelihood-ratio statistic tests exogeneity of infant morbidity, arm circumference and length; degrees of freedom = 6, 12 and 18, respectively.
*P <0.05.
Source: Bhargava (2000).

months can shed useful light for designing food policy interventions. This is especially true when data on children's food intakes are assessed, as in the case of Kenyan infants. The results from estimating a dynamic random-effects model for the length, head circumference and weight of 102 Kenyan infants observed at monthly intervals in six time periods are presented in Table 4.2. Four indicator variables are included in the model for time periods 3–6, and infants' anthropometric measurements in the previous month were explanatory variables in the respective models. The main findings in Table 4.2 were, first, that boys were taller than girls. This was in contrast with the results in Table 4.1, where gender differences in length at birth were not statistically significant. Second, infants' intakes of calcium through weaning were significant predictors of length. This is an important finding and similar results were reported for heights of school-aged children in this population (Bhargava, 1999). Thus, increasing the intakes of dairy products is likely to enhance children's linear growth in Kenya.

Third, the indicator variables for time periods 3–6 were estimated with positive coefficients that were statistically significant, showing a steady increase in length over time. The lagged dependent variable was estimated with the coefficient 0.36, which was significant. The between–within variance ratio was estimated to be 0.56, indicating that there remained significant unobserved between-children differences in the data. Last, while the coefficient of the morbidity index was estimated with a negative sign, it was not statistically significant at the 5% level. Moreover, the likelihood ratio test accepted the exogeneity null hypothesis that the random effects affecting length were uncorrelated with the errors affecting the children's morbidity index. This was perhaps not surprising in view of the statistical insignificance of the morbidity index in the model for infant length.

The results for the head circumference of Kenyan infants are also presented in Table 4.2. The results showed that boys had significantly larger head circumference than girls, as was also seen for the measurements at birth (Table 4.1). Second, maternal hemoglobin concentration was a significant predictor of infant head circumference. This finding reflects the beneficial effects of good maternal iron status for lactation and other activities. Third, maternal head circumference was a significant predictor of infant head circumference. While this association may reflect the influence of genetic factors, paternal head circumference was not a significant predictor of infant head circumference. Fourth, maternal BMI and infant morbidity were not significantly associated with infant head circumference. Indicator variables for time periods were estimated with significant coefficients, and the coefficient of the lagged dependent variable was 0.30. Fifth, the between–within variance ratio was estimated to be close to unity, indicating significant unobserved between-children differences in the data. Last, infant length was a significant predictor of head circumference, presumably because growth in length can increase cranial vault thickness (Lieberman, 1996). Taking into account such information from the physical anthropology literature improved the specification of the model and ensured robust inferences.

The results for infants' weights in Table 4.2 showed the importance of maternal hemoglobin concentration, which was estimated with a positive and significant coefficient. Maternal BMI was a significant predictor of infant weight, though paternal BMI was not significant. The infant morbidity index was estimated with a negative sign and was significant. Moreover, arm circumference and length were estimated with large positive coefficients, which were significant. As discussed in Chapter 3, length approximates skeletal size, while arm circumference reflects muscle mass; it was not surprising that these two variables were important predictors of infants' weights. The lagged dependent variable was estimated to be 0.28 and was significant. In contrast with the results for infant length and head circumference, the between–within variance ratio was not significant in the model for weight. While unobserved differences in length and head circumference were likely to be influenced by genetic factors, genetic factors may be less important for explaining weight, since weight, responds to environmental factors in a shorter time frame.

In summary, the results for infant length, head circumference and weight showed beneficial effects of maternal nutritional status, infants' nutritional intakes and morbidity for physical growth. Broadly speaking, intra-uterine growth and infant growth in the early months are complex processes that are poorly understood. Thus, for example, Indonesian women who received a high-energy supplementation in the last trimester of pregnancy did not produce infants that were heavier at birth than women receiving a low-energy supplement (Kusin *et al.*, 1992). Instead, infants of women receiving high-energy supplement were heavier during the period 3–24 months and were taller in the period 3–60 months. Thus, maternal nutritional status is likely to affect infant growth with complex time lags. While food intakes by Kenyan women during pregnancy were not significant predictors of infants' anthropometric indicators, anemic and low-BMI women are important target groups for receiving nutritious food supplements. For example, increasing the intake of animal products containing high quantities of absorbable iron by mothers would be an effective strategy for enhancing the growth of infants in the early months.

4.3 Growth of pre-school children in developing countries

Children's physical development

The physical growth of pre-school children has been the focus of many studies in developing countries by researchers in the fields of anthropometric assessment and nutrition (e.g. Tanner, 1966, 1986; Waterlow *et al.*, 1977; Martorell and Habicht, 1986). As noted in equation (3.1), the z-scores for a child's height-for-age (or weight-for-age) are computed with reference to heights (or weights) of children in the same age-groups in the US (NCHS, 1977) or UK (Freeman *et al.*, 1995). An advantage in using z-scores is that one can compare growth patterns over time of children in developing countries with those in the reference population. Comparisons across different developing countries are also possible though, as noted in Chapter 5, caution is needed in applying z-scores to the investigation of gender differences.

A useful application of z-scores for heights of boys and girls in Honduras was presented by Martorell and Habicht (1986, fig. 5). At noted in section 4.2, length at birth of children in developing and developed countries are similar. However, it was evident for Honduran children that the z-scores of heights started to decline rapidly with age. For example, by the time Honduran children were 24 months old, their heights were two standard deviations below the heights of children in the US. Moreover, the heights of Honduran boys at 7 years were more than 2.5 standard deviations below the US heights. Nutritional intakes and environmental factors such as disease due to poor sanitation are likely to contribute to differences in children's heights. Moreover, intra-uterine growth retardation can induce hormonal changes; it might be difficult for children in developing countries to reach their full growth potential even if the environment were drastically improved.

Because growth retardation is common in developing countries, nutritionists have conducted several food supplementation studies to quantify the effects of nutrient deficiencies on child growth. A well-known study was conducted by the Institute of Nutrition of Central America and Panama (INCAP) in the period 1968–77; a follow-up was conducted in 1988/9

(Martorell and Scrimshaw, 1995). Briefly, an intervention in certain Guatemalan villages offered a high-protein food supplement, "Atole", to pregnant and lactating mothers and their children below 7 years of age. In the control villages, a drink, "Fresco", was offered which contained no protein but was fortified with nutrients such as iron and vitamins A and C. The energy content of Atole was 119 kcal, whereas that of Fresco was approximately half (59 kcal). One of the emphases in the INCAP study was the effects of higher protein intakes on brain development and consequently on children's scores on cognitive tests. The average growth in length of 675 children in the age group 3–36 months receiving Atole was significantly higher than that for children receiving Fresco (Schroeder *et al.*, 1995). Similarly, average growth in weight for 454 children receiving Atole was significantly higher in the age group 3–24 months than for the Fresco group.

Furthermore, follow-up of supplemented children a decade later in 1988/9 indicated that the females who received Atole were significantly taller and heavier as adolescents (Rivera *et al.*, 1995). In another study by Haas *et al.* (1995), adolescents' physical work capacity was assessed by measuring the volume of oxygen consumed; higher oxygen uptake reflects greater physical work capacity. Certain multivariate regression models for males showed significant beneficial effects of Atole supplementation on physical work capacity. These findings were not supported for females, where the indicator variable in the model for Atole supplementation was insignificant. However, physical activity patterns of men and women are often different (Chapter 6); better-nourished females are likely to produce healthier infants that show higher growth in the early months (see section 4.1). Moreover, investigators analyzing the INCAP follow-up data did not exploit some longitudinal features of the data. For example, including the information in the model on the age at which children showed signs of "stunting" or "wasting" in the earlier part of the study (1968–77) may have provided insights into the likely benefits of food supplementation on the physical work capacity of adolescents.

Lastly, while there are numerous studies investigating the effects of food supplementation on the growth patterns of pre-school children, sample sizes are usually small due to the costs of collecting high-quality information on nutritional and biological variables. For example, a collection of papers in Pollitt and Schurch (2000) investigated the effects of energy and micronutrient supplementation on the growth of Indonesian children in the age group 12–24 months; the number of children in the various intervention subgroups was around 20. Because there are numerous links between nutrition and child growth, such studies are likely to show beneficial effects of improved nutrition on child outcomes. However, the small sample sizes limit the scope of the analyses. If pre-school children show signs of energy and nutrient deficiencies, then food policies alleviating the deficiencies are likely to be beneficial. The design of food policies would benefit from quantitative analyses covering a large number of children together with socio-economic information on the households. While such studies are expensive, allocation of resources between interventions for pregnant mothers, and pre-school and school-aged children, is likely to benefit from the results of thorough analyses. However, while piecemeal approaches to data on children in different age groups are inevitable due to budget constraints, these approaches provide many useful policy insights.

Cognitive development of pre-school children

The cognitive development of infants and pre-school children in developing countries has been investigated by psychologists and nutritionists (e.g. Lozoff, 1988; Grantham-McGregor, 1995). Infants born in developed countries are often tested on the Brazelton Neonatal Behavioral Assessment Scale (Brazelton, 1984) and this scale has also been used in developing countries. For example, in the Kenyan study in the Embu region, infants were tested using the Brazelton scale. Empirical analyses, however, indicate that the Brazelton scale is unlikely to identify malnourished children unless the infants suffer from neurological defects (Bhargava, 2000). Because children's physical and cognitive developments are intertwined at low levels of nutrition, it is useful to assess children's cognitive development from early ages. Thus, for example, the Bayley Motor Scale and the Bayley Infant Behavior Record (Bayley, 1984) contain several items that can be administered to children around the age of 6 months. However, the disadvantage in giving psychological tests to very young children is that the tests may be unreliabile, e.g. children may score differently depending on their mood or on chance. The Brazelton scale is a good example of low reliability, though Bayley tests at 6 months can be useful in many circumstances. Moreover, psychological tests require detailed input from researchers and need to be adapted to the cultural setting. By contrast, anthropometric variables assessing physical growth are relatively inexpensive to measure.

Further, for pre-school children in the age group 2–3 years, the Bayley Motor and Mental Scales are useful for assessing development (Bayley,

1969). For example, Sigman *et al.* (1989) reported positive correlations between Kenyan children's length and weight and the scores on the Bayley Mental and Motor Scales. Moreover, the scores on the subcomponents "Symbolic play" and "Amount of verbalization" were positively correlated with children's protein intakes from animal sources. Although diet quality was correlated with some of these items, correlations between animal protein intakes and overall scores on the Bayley Mental and Motor Scales were not significantly different from zero. Using a multivariate framework, Sigman *et al.* (1989) found that variables such as children's weight, amount of verbalization, and social interactions were significant predictors of the scores on the Bayley Mental Scale at 30 months.

Overall, results from the Kenyan study indicated that psychological tests can be useful for assessing the development of young children, though it is important to collect anthropometric data as well. Unlike in developed countries, pre-school children growing up in illiterate households in developing countries may not be exposed to early stimulation that translates into better cognitive outcomes. Thus, even if pre-school children perform poorly on psychological tests at (say) 3 years of age, such children can do well in school, especially if the learning environment is stimulating. Of course, if pre-school children are stunted or wasted, then policy-makers should tackle these problems via food supplementation programs. Because children's performance on school examinations is often a deciding factor for their ability to complete primary education, the determinants of children's scores on various tests are analyzed in Section 4.4.

4.4 Cognitive development of school-aged children: theoretical considerations

As discussed in sections 4.1–4.3, nutrient deficiencies can adversely affect children's brain development. While in the last four decades considerable effort was spent investigating the effects of protein deficiencies, current thinking underscores the role of micronutrient deficiencies, especially of iron (e.g. Beard, 1995; Scrimshaw, 1996). However, under-nutrition and frequent spells of hunger are likely to affect not only the central nervous system but also children's overall psychological development. In order to assess the full consequences of under-nutrition, therefore, it is helpful to examine certain theories of early child development. Quantitative analyses incorporating the conceptual aspects will be useful for understanding the effects of under-nutrition and the home and school environments on the cognitive development of school-aged children.

Some aspects of Vygotsky's theories of speech, learning and development

During the 1930s, L. S. Vygotsky emphasized that speech plays an important role in the development of very young children (Vygotsky, 1987). The child attempts to convey its needs to adults in the environment with the help of seemingly "egocentric" speech. However, as the child masters words and undergoes other developmental processes, speech is gradually internalized in that it is no longer necessary to speak aloud while performing tasks. This "inner" speech facilitates the organization of the child's mental functions, while aiding the task of communicating with others. This approach contrasts with earlier work by Piaget (1977) attributing early speech chiefly to biological factors. It is argued below that Vygotsky's approach is more suitable for interpreting cognitive data, especially for children growing up in poverty.

In other chapters of his book, Vygotsky distinguished between "spontaneous" concepts, which are learnt from the child's practical activity, and "scientific" concepts, acquired through instruction in school or at home. The development of spontaneous concepts was argued to proceed from specific to more abstract form, whereas the converse was the case for scientific concepts. In spite of the fact that the two processes move in opposite directions, the connections between them were underscored and the boundary between them was argued to be a fluid one. Moreover, understanding scientific concepts exerts pressure on a child's thought processes, thereby increasing awareness and placing the future development of spontaneous concepts on a higher plane of maturity.

Last, since the development of scientific concepts is intertwined with instruction, Vygotsky considered dynamic interactions between learning and development. Learning was argued to lead, over time, to child development by spurring mental processes that in turn enhance development. Furthermore, the distance between a child's actual development and the level that can be achieved with help from an adult was defined as the "zone of proximal development". This was a dynamic concept, since the zone of proximal development changes throughout children's development. Some implications of this concept in mostly non-quantitative settings were explored by Rogoff and Wertsch (1984).

Differential effects of under-nutrition on the components of cognitive tests

Children growing up in poverty face under-nutrition, poor stimulation from the environment, and an inadequate educational infrastructure. While it may seem complicated to assess the contribution of all these factors to children's development, analyses including variables reflecting the diverse aspects are useful from a policy standpoint. Furthermore, the development of speech occurs at the earliest stages and speech serves the vital function of enabling the child to express its needs (Vygotsky, 1987); interaction with adults and peers enhances external speech. Thus, very young children may be able to attain their potential with respect to verbal abilities in as much as their basic mental functions are not compromised by chronic food shortages and/or infections. The development of inner and written speech, however, demands greater abstraction. If certain parts of a child's brain are not well developed due to nutritional deficiencies, then in spite of the known plasticity of the brain (Purves, 1988), the child will be slow in realizing its potential on tasks involving higher mental functions (Binet and Simon, 1916). However, nutrient deficiencies beyond a certain level can compromise the capacity to learn abstract concepts. Thus, scores on different components of cognitive tests are likely to be differentially affected by indicators of nutritional status and other factors. For example, brain development might play a greater role in the scores achieved on quantitative and analogical reasoning. Because these test scores also depend on the instruction process in school, it is necessary to consider the dynamics of the zone of proximal development for children in developing countries.

Under-nutrition and the dynamics of the zone of proximal development

The zone of proximal development, is an important concept for understanding children's development, though its measurement is complicated. However, this distance must diminish with age as children attain their potential. For children living in poverty, both nutritional status and learning affect this zone. It would be helpful to consider three broad cases that are applicable to children in developing countries. First, if practically all children face chronic energy and micronutrient deficiencies and the learning environment is extremely poor, then the zone of proximal development will be quite narrow. Children will not grasp many concepts in school and scores on cognitive tests will provide few insights because between-children differences will be small. In such circumstances, it is unlikely that children will pursue school beyond a few primary grades, and this is often the case in many poor countries, especially in sub-Saharan Africa.

Second, if some children are adequately nourished while others suffer from nutritional deficiencies and the learning environment is moderately stimulating, then one will observe significant between-children differences in scores, especially on tasks involving higher-order mental functions. Moreover, these differences will be systematically related to indicators of nutritional status and environmental factors. For under-nourished children, the zone of proximal development will begin shrinking at an earlier age and the situation will deteriorate over time; as adolescents, such children are likely to drop out of school as the concepts become more abstract. This seems to be the situation in many developing countries. Third, if the children are well nourished but the school and home environments are not very stimulating, then nutritional status may not be a good predictor of test scores. In developed countries such as the US, many under-privileged children face such constraints and their cognitive development is often compromised by a lack of opportunities for adults in the household.

Specification of models for scores on cognitive tests and school examinations

Learning and mental development are dynamic processes since their future realizations depend on the knowledge accumulated by the child. Moreover, continuous interactions between learning and development determine the "equilibrium" level of cognitive development. At a given point in time, scores on cognitive tests and school examinations reflect certain aspects of children's mental capabilities and knowledge. For example, scores on quantitative components will be influenced by the ability to accurately recall and apply the appropriate concepts. Because such abilities result from complex biological interactions that involve nutritional and other unknown mechanisms (Levitsky and Strupp, 1995), models for test scores and school examinations should incorporate child characteristics and account for unobserved between-children differences.

While in developed countries one might not observe a relationship between test scores and indicators of nutritional status (Southon *et al.*, 1994), scores of children in developing countries are likely to depend on their nutritional status and on home and school environments. Children's nutritional status can be approximated by their height, weight, head and arm circumferences, hemoglobin concentration, morbidity and nutrient intakes. Home environment is reflected in parental education and socio-economic and demographic variables. School infrastructure is often difficult to approximate, though classroom observations on books, blackboards, desks, assignments, etc., for children, as well as teachers' qualifications, are useful indicators. Also, one needs to control for the grade level of the child, since this reflects past learning (Ceci, 1991).

Alternative measures for schoolchildren's cognitive development

It is common in psychological research to give a battery of tests to children in developing countries in order to assess various dimensions of cognitive development (e.g. Neumann *et al.*, 1992; Pollitt *et al.*, 1993). This is useful in part because the reliability of tests can be poor and aggregate test scores are often better indicators of child development. Furthermore, children's scores on school examinations in different subjects can be easily compiled from class registers. While scores on school examinations are good indicators of learning in developed countries (Goldstein and Thomas, 1997), this may not be the case in developing countries where educational infrastructure is poor; teachers may assign grades (marks) based on classroom participation rather than on learning. Thus, tests devised by researchers in education and psychology and administered by external enumerators can provide a better assessment of children's learning in developing countries.

In a related vein, certain psychologists have suggested that intestinal parasites, such as hookworm, in developing countries can impair children's cognitive function via their effects on the central nervous system (e.g. Watkins and Pollitt, 1997); children have been given several cognitive tests to explore this hypothesis. For example, schoolchildren in Tanzania were given tests such as the "grooved pegboard", where the time taken to complete the task reflects children's coordination abilities (Partnership for Child Development, 2002). Psychologists administer tests such as "digit span", where the child repeats numbers in forward and backward orders. Also, items from Raven's Progressive Matrices (Raven, 1967) which entail identifying the correct box are used, because these tests are neutral with respect to language and culture. In addition, children are given tests in spelling, reading and arithmetic, and these are often referred to as "educational achievement tests" since they are closer to the material taught in the class. Overall, scores on cognitive and educational achievement tests are likely to capture different dimensions of child development, and analyses of the data can provide useful insights. Because there have been many interventions for removing intestinal parasites in developing countries, the importance of a comprehensive modeling approach to the data emerging from randomized trials is discussed in the next subsection.

Children's cognitive development and the data from randomized trials

Ever since the pioneering work by Fisher (1935), the use of randomized trials has been common in agricultural and biomedical research. Several study designs are covered in standard statistical texts (e.g. Cox, 1958). From the standpoint of assessing the relative importance of various factors affecting child development in developing countries, randomized trials have investigated possible benefits on cognitive function of removing intestinal parasites such as hookworm. This literature was reviewed by Dixon *et al.* (2000), who argued that on the basis of published findings, it was not worthwhile to de-worm children. Contrary opinions were published in two subsequent issue, of the *British Medical Journal* (Dickson *et al.*, 2000) and, as discussed below, a broader approach to data analysis is necessary for resolving differences in opinions.

Investigators often assess the benefits of antihelmintic treatment on children's cognitive test scores by comparing changes in treatment and control groups; independent t-tests are used to determine whether differences in changes in the two groups are statistically significant. As emphasized above, children's cognitive test scores are influenced by a wide range of factors, including their nutritional and health status, and home and school environments. Moreover, de-worming is likely directly to affect children's hemoglobin and ferritin concentrations, since intestinal parasites thrive on blood. By contrast, children's cognitive test scores can improve due to de-worming because of overall improvement in health status which, in turn, can increase effective participation in school. In order to understand the role of factors affecting children's cognitive development, therefore, researchers need to model the relationships between health status, environmental factors and cognitive outcomes. Econometric analyses of the pathways can shed light on the role played by various factors.

Further, analyses of data from the control group of randomized trials of de-worming are useful for understanding the determinants of child health and cognitive development. Moreover, empirical models for the control group can be used to predict changes that are likely to be observed in the treatment group. For example, Bhargava, Jukes *et al.* (2003) used the data from a randomized trial removing hookworm and schistosomiasis infections among heavily infected Tanzanian schoolchildren. Children in the two groups were observed at baseline, 3 and 18 months, with highly infected children in the treatment group receiving chemotherapy (albendazole for three days against hookworm and/or a single doze of praziquantel against schistosomiasis) at 3 and 18 months. Random-effects models were estimated for children's hemoglobin and ferritin concentrations using data from the control group; it was predicted that reducing hookworm and schistosomiasis loads to zero would lead to an increase in children's hemoglobin concentration of approximately 3%. However, data from the treatment group showed an increase of 8% among the treated children, i.e. the longitudinal random-effects models significantly under-predicted the benefits of de-worming. These findings could be due to improved absorption of nutrients for children who became free of intestinal parasites. This comprehensive modeling approach to the data from control and treatment groups underscored the importance of jointly addressing biological and statistical issues. Integrated analyses are likely to facilitate policy formulation in situations where there is a strong interface between biological and socio-economic issues. The results from empirical models for Kenyan and Tanzanian schoolchildren's scores on various cognitive and educational achievement tests are discussed next.

4.5 Cognitive development of school-aged children: results from Kenya and Tanzania

Empirical results for Kenyan children's scores on cognitive tests and school examinations

Table 4.3 presents the results for Kenyan children's scores on several cognitive tests, using the data from the Embu region of Kenya at three time points; the results for subcomponents digit span, Raven's matrices and arithmetic, word meaning, and behavioral cooperation are also presented (Bhargava, 1998). The scores on digit span, Raven's matrices and arithmetic are likely to reflect the children's analytical abilities or their "higher order mental functions" (Binet and Simon, 1916). By contrast, behavioral cooperation consisted of scores assigned by the enumerators on tasks that reflected the child's cooperativeness, goal directedness, attention span and judgment of the test; these scores may be influenced by enumerator bias. The models controlled for children's nutritional and health status, morbidity, grade level ("standard"), socio-economic

Table 4.3 Maximum-likelihood estimates of models for Kenyan children's scores on cognitive tests

| | Dependent variable | | | | | | |
| | Total score | | Digit, span + Raven's matrices + Arithmetic | | Word meaning | | Behavioral cooperation[a] | |
Independent variable	Coefficient	SE	Coefficient	SE	Coefficient	SE	Coefficient	SE
Constant	1.077	0.665	6.523	3.046	0.456	2.723	0.784	2.330
Hemoglobin	0.089	0.104	0.234*	0.108	0.098	0.149	−0.022	0.130
Parents' scores[b]	0.108	0.065	0.151*	0.075	0.135	0.140	0.178	0.095
Standard	0.169*	0.038	0.279*	0.061	0.231*	0.057	0.085*	0.032
Socio-economic status + cash income[c]	0.082*	0.042	0.057	0.078	0.197*	0.064	−0.018	0.050
Head circumference	0.647*	0.037	1.658*	0.856	0.029	0.664	0.322	0.548
Body mass index	0.351*	0.166	0.410	0.231	0.698*	0.261	0.146	0.205
Morbidity[d]	−0.014*	0.006	−0.023*	0.010	−0.006	0.009	−0.015	0.009
Lagged dependent variable	0.191	0.112	0.083	0.105	0.151	0.148	0.037	0.092
Between–within variance	0.412	0.264	0.779*	0.379	0.382	0.311	0.072	0.09
Within variance	0.012		0.031		0.026		0.029	
Chi-square (9)[e]	9.80		17.87		0.65		12.40	
Sample Size (n)	104		104		104		106	

Note: All variables in logarithms; children were observed in 3 survey rounds.

[a] Consists of cooperativeness, goal directedness, attention span and judgment of the test.

[b] Parents' scores in the four rightmost columns correspond to the respective components for children.

[c] Socio-economic status is based on household possessions.

[d] Based on number of days the child was sick.

[e] A likelihood-ratio test statistic for exogeneity of head circumference, BMI and morbidity.

*P.<0.05.

Source: Bhargava (1998).

variables and parental scores on cognitive tests. The models were dynamic in that the previous score was an explanatory variable for explaining scores in the current period. The parameters were estimated by the max-imum-likelihood method and between and within variances were esti-mated using the simple random-effects decomposition.

The main findings in Table 4.3 were as follows. First, children's BMI and head circumference were significantly positively associated with overall scores, and morbidity was negatively associated. Thus, measures of chil-dren's nutritional and health status were important for explaining their cognitive development. Second, while hemoglobin concentration was not a significant predictor of overall scores, parental scores on cognitive tests, household socio-economic status and child's grade level were. The lagged dependent variable was significant and so was the between–within vari-ance ratio, indicating that there remained significant unobserved be-tween-children differences. Third, in the model for scores on digit span, Raven's matrices and arithmetic, the coefficient of the head circumference was large and significant. As noted previously, head circumference reflects brain mass; in this homogeneous ethnic population from Kenya, head circumference was a useful predictor of children's cognitive development. Children's BMI and hemoglobin concentration were significant predictors of the scores on digit span, Raven's matrices and arithmetic, while mor-bidity was significantly negatively associated. However, the lagged depen-dent variable was not significant in this model, in part due to disaggregation of the scores.

Fourth, in the model for scores on word meaning, child's grade level and BMI and socio-economic status of the household were significant predict-ors. As discussed in section 4.4 in the context of Vygotsky's theories, a child must learn language to communicate with adults in the immediate environment. Thus, the effects of nutritional and health status were more likely to be visible in the scores on tests reflecting analytical abilities. Lastly, in the model for scores on behavioral cooperation, only the child's grade level and the household's socio-economic status were significant predictors. These scores were likely to be influenced by the subjective assessments of enumerators, though they can be useful for assessing the extent to which children were eager to learn in the school environment.

Furthermore, results for the scores of Kenyan children on school exam-inations showed that variables reflecting nutritional and health status were not significant predictors (Bhargava, 1998). Significant variables in these models were child's grade level and number of days the child was absent from school. Thus, children attending school more regularly scored higher on school examinations. By contrast, the absent variable was not a significant predictor of cognitive test scores in Table 4.3. Taken together, these results suggested that the school environment in the Embu region was not very stimulating, though cognitive tests identified health and environmental factors affecting child development. Thus, it is important in impoverished settings to have independent assessments of children's learning to facilitate the channeling of resources for improving child health and educational outcomes.

Empirical results for Tanzanian children's school attendance and scores on cognitive and educational achievement tests and school examinations

The data on cognitive test scores of children in the control group of the randomized trial in Tanzania removing hookworm and schistosomiasis were modeled by Bhargava, Jukes *et al.* (2005). A notable feature of this trial was that the children were given a battery of cognitive tests designed by psychologists to assess the possible effects of intestinal parasites on cognitive function. Moreover, children took educational achievement tests in spelling, reading and arithmetic; data on marks obtained on school examinations in four subjects were also recorded. However, changes in scores on various tests between baseline and 18 months were not significantly different for children in the control and treatment groups. Thus, if one focused only on the hypothesis driving the randomized trial, one would conclude that de-worming did not affect children's cognitive performance. However, this would be a narrow interpretation of the effects of de-worming because one mainly expects a direct relationship between de-worming and indicators of iron status such as hemoglobin and ferritin concentrations (Bhargava, Jukes *et al.*, 2003); hemoglobin concentration was an important predictor of cognitive test scores of Kenyan children. Thus, the data from the control group enable modeling of the pathways underlying children's cognitive development, which depend on several socio-economic factors.

Bhargava, Jukes *et al.* (2005) presented analyses of the data from the control group on children's school attendance, and scores on cognitive tests, educational achievement tests and school examinations. Note that children's C-reactive protein levels were included in the models because these rise with inflammations and reflect morbidity. Also, children's school attendance was excluded from the models for scores on cognitive tests because it was not a statistically significant predictor. Last, the models for scores on educational achievement tests were estimated taking into account teachers' experience and number of work assignments in the class. A variable interacting children's heights with work assignments was included as an explanatory variable in these models to investigate multiplicative effects.

The main results from estimating the empirical models can be summarized as follows. First, in the model for school attendance, households' socio-economic status was estimated with a positive and significant coefficient. Thus, children from better-off households attended school more regularly. Moreover, children's hookworm and schistosomiasis loads were not significant predictors in the model for school attendance. Because poverty is a critical factor in hampering children's continuation in school, regular school attendance is likely to be facilitated via subsidies to poor households. This is especially important in the wake of the AIDS epidemic in Africa, where children are being drawn into unpaid and paid tasks due to parental deaths (Bhargava, 2005).

Second, the results in Table 4.4 for children's scores on digit span, Corsi block, Stroop and Pegboard time using the dominant hand showed that taller children did better on these tests. However, hookworm and schistosomiasis loads were insignificant predictors; schistosomiasis load had an unexpected negative sign in the model for Pegboard time using the dominant hand. These results partly supported the findings from the treatment group that de-worming did not significantly improve children's scores on cognitive tests. Also, household socio-economic status was not a significant predictor of the scores on cognitive tests. Partly because of the difficulties in designing culture-specific cognitive tests and also because of the difficulties in explaining some of the tasks, it appears that cognitive tests may not be the most appropriate indicators of child development in poor countries. Of course, in situations where specific cognitive tests accurately reflect the effects of underlying biological processes, improvement in child health might have beneficial effects on the scores for such tasks.

Third, in contrast with cognitive tests, children were familiar with the educational achievement tests that were administered in the Tanzanian study. The results in Table 4.5 show the importance of grade level, height, hemoglobin concentration and school attendance for the scores on sentence reading, arithmetic and spelling. Hookworm and schistosomiasis loads were not significant predictors, indicating that de-worming in the treatment group was unlikely significantly to improve children's performance on these tests. This, however, does not imply that de-worming was not beneficial for children's health; de-worming significantly increased children's hemoglobin and ferritin concentrations (Bhargava, Jukes *et al.*,

Table 4.4 Efficient estimates from the random-effects models for Tanzanian children's scores on cognitive tests in 3 survey rounds, explained by socio-economic and anthropometric variables, infections and hemoglobin concentration

	Dependent variable							
	Digit span (forward + backward)		Corsi block		Stroop (forward + backward)		Pegboard dominant time	
Independent variable	Coefficient	SE	Coefficient	SE	Coefficient	SE	Coefficient	SE
Constant	-2.654*	0.997	-1.000	0.816	9.820*	0.719	9.717*	0.982
Grade	0.034*	0.012	0.018*	0.009	-0.042*	0.008	-0.034*	0.013
Socio-economic status	-0.062	0.136	-0.102	0.102	-0.110	0.102	-0.227	0.144
Hookworm (eggs/g)	0.0004	0.004	-0.003	0.003	-0.002	0.003	-0.004	0.004
Schistosomiasis (eggs/10 ml)	0.003	0.004	0.004	0.003	0.0007	0.003	-0.013*	0.005
Height (m)	1.048*	0.166	0.691*	0.139	-1.058*	0.122	-0.724*	0.164
C-reactive protein (mg/dl)	-0.010*	0.005	-0.003	0.005	0.004	0.003	0.008*	0.004
Hemoglobin (g/1)	0.023	0.072	0.064	0.063	-0.080	0.051	-0.168*	0.068

Note: Values are slope coefficients and standard errors (n = 359, t = 3). Except for Grade, all variables were in natural logarithms; coefficients of independent variables in logarithms were the elasticities. The coefficients of school indicator variables and school attendance were not significant and were therefore dropped from the models.
*P < 0.05.

Source: Bhargava, Jukes *et al.* (2005).

Table 4.5 Efficient estimates from the random-effects models for Tanzanian children's scores on educational tests in 2 survey rounds, explained by socio-economic and anthropometric variables, infections, hemoglobin concentration, school attendance and interactions between health status and school infrastructure

| | Dependent variable | | | | | |
| | Sentence reading | | Total arithmetic | | Spelling | |
	Coefficient	SE	Coefficient	SE	Coefficient	SE
Independent variable						
Constant	−24.143*	4.041	−0.978	1.218	−15.216*	3.188
Grade	0.198*	0.032	0.110*	0.010	0.132*	0.028
Socio-economic status	0.269	0.373	0.094	0.114	0.742*	0.341
Teacher experience	0.026	0.017	0.013*	0.005	0.012	0.016
Work assignments	0.212*	0.100	−0.037	0.030	0.044	0.078
Hookworm (eggs/g)	−0.007	0.010	−0.001	0.002	−0.009	0.009
Schistosomiasis (eggs/10 ml)	0.005	0.011	−0.0002	0.003	0.006	0.010
Height (m)	4.717*	0.737	0.510*	0.225	2.786*	0.575
Work assignments × Height	−0.042*	0.020	0.008	0.006	−0.009	0.016
C-reactive protein (mg/dl)	0.016	0.017	0.001	0.005	0.006	0.011
Hemoglobin (g/l)	0.420*	0.201	0.154*	0.059	0.264	0.150
Proportion of school attendance	0.033*	0.012	0.020*	0.006	0.033*	0.008

Note: Values are slope coefficients and standard errors (n = 507, t = 2). Except for Grade, all variables were in natural logarithms; coefficients of independent variables in logarithms were the elasticities. School indicator variables were omitted from these models. Chi-square statistics were suppressed.
*P <0.05.
Source: Bhargava *et al.* (2005).

2003). Furthermore, the models showed the importance of teacher experience and work assignments for the scores on educational achievement tests. The scores on sentence reading showed the presence of interactive effects between children's heights and work assignments. Thus, additional work assignments can improve children's learning, though there are limits to the workload in view of the prevalence of under-nutrition among the Tanzanian schoolchildren. Moreover, larger sample sizes would be necessary to estimate precisely the threshold points in non-linear models that can shed light on the potential benefits of increases in work assignments depending on a child's height-for-age.

Lastly, the results for Tanzanian children's scores on school examinations in arithmetic, science, geography and physics were not very informative (Bhargava, Jukes *et al.*, 2005). Most of the explanatory variables were not significant predictors and, unlike the results for the Kenyan schoolchildren (Bhargava, 1998), even school attendance was not a significant predictor. While scores on school examinations are inexpensive to compile, it appears that poor educational infrastructure in developing countries is likely to diminish their usefulness. Moreover, educational achievement tests are easy to design and the marginal costs of implementing them entail the hiring of qualified local enumerators. Thus, educational achievement tests appear to be the most suitable indicators of child development in impoverished settings. Moreover, the results for cognitive test scores in the Kenyan study (Table 4.3) apparently benefited from inclusion of components such as children's scores on arithmetic, which is a component of educational achievement tests. Of course, the availability of data on scores on cognitive and educational achievement tests and school examinations in the Tanzanian study enabled a systematic comparison of the alternatives for assessing children's learning and development in developing countries.

4.6 Conclusions

The purpose of this chapter was to review broadly the literature on child development in developing countries from infancy to primary education. There are numerous studies in biomedical fields and it is important for social scientists to understand the motivation for compiling data on different indicators of child development. For example, it was argued in section 4.1 that anthropometric indicators such as length, head circumference and weight are likely to be more useful than psychological indicators for assessing infant development. This may also be true for pre-school children, though Bayley-type tests can be administered to assess cognitive development. While various constructs have been developed in the psychological literature, economists are generally more concerned with the proximate determinants of children's physical and mental development. The two approaches generally converge for school-aged children, where a battery of tests has been given to children in many developing countries. Moreover, Vygotsky's theories of child development, discussed in section 4.4, are useful for the specification of empirical models for children growing up in poverty.

The results from the analyses of data from Kenya and Tanzania in section 4.5 showed the importance of children's health status for their scores on cognitive and educational achievement tests. The empirical results identified key determinants of child development. For example, the index of the socio-economic status of households and biological variables such as height, weight, head circumference and hemoglobin concentration were important predictors of scores on cognitive tests. Furthermore, the importance of randomized controlled trials in developing countries was outlined and suitable interpretation of the data from such studies was emphasized in the discussion. It is important for social scientists to appreciate that specific biological hypotheses drive randomized controlled trials, though the emerging data can be analyzed in a broad framework. For example, the benefits of de-worming children are likely to be confined to improvements in iron status, which in turn, may have beneficial effects for child development. However, de-worming in itself is unlikely to increase school attendance except on days of testing and distribution of medicines, when parents may see additional benefits of sending children to school. Moreover, the AIDS epidemic in sub-Saharan Africa is reducing school attendance, especially of orphaned children; children's cognitive development is therefore likely to benefit from subsidies for education (Bhargava and Bigombe, 2003; Bhargava, 2005). Overall, multi-disciplinary analyses are useful for the formulation of health and education policies, since they take into account the diverse factors affecting child development. Policies based on results from such analyses are likely to increase the future supply of skilled labor and promote economic growth in developing countries.

5

Fertility, child mortality and economic development

The issues of high fertility (i.e. a large number of children born to women) and child mortality are a major focus of demographic research in developing countries. These aspects are critical for economic development because the health status of surviving children and the ability of poor households to educate them depend on family resources that dwindle with additional births. Moreover, high fertility rates and low employment prospects in rural areas of developing countries increase the migration of unskilled labor to urban areas, which in turn leads to congestion and slums. However, demographic aspects are seldom well integrated in nutritional and economic research. This may be partly due to the nature of demographic surveys, which compile information on large numbers of married women in developing countries for investigating the history of births and patterns of child deaths; there is less emphasis on economic aspects and variables reflecting health status are not measured with the precision that biomedical scientists are accustomed to. For example, the National Family Health Survey in 1992/3 (NFHS-1) covered 89,000 Indian women in the age group 13–49 years (International Institute for Population Sciences, 1995). Because child mortality is the most extreme form of morbidity, it is natural to view the determinants of child mortality as an extension of factors affecting child morbidity (analysed in Chapter 3). Moreover, children that die before reaching the age of 1 year (infant mortality) or before the age of 5 years (under-5 mortality) are often under-nourished and sick prior to death. Several children born to a poor woman within a short time interval reduces the resources available for food and health care. Thus, factors underlying high fertility rates are important for understanding the determinants of children's nutritional and health status and cognitive development. The links between fertility, children's nutritional status and cognitive development have perhaps not been clearly delineated in demographic research (Bhargava, 2001a).

At least two sets of issues relating to child health have been emphasized in the economic demography literature. First, it has been argued that girls face discrimination in developing countries, such as those in south Asia. Second, female education has been emphasized as an important instrument for lowering child mortality and fertility rates. While female education increases maternal awareness of hygiene and vaccinations, the mechanisms through which female education lowers child mortality merit closer consideration. For example, countries such as China have afforded easy access to health care in rural areas in the last few decades; consequently, there has been a sharp decline in child mortality even in regions with low female education levels. Of course, female education enhances children's cognitive development (see Chapter 4), and the benefits of maternal education are wide-ranging and should be discussed in specific contexts.

Further, economic and demographic research has utilized data at the national, regional and household levels to draw policy implications. In this chapter, the diverse literature is briefly outlined to highlight the importance of demographic factors for economic development. Section 5.1 begins by pointing out methodological difficulties in using anthropometric indicators such as children's heights and weights for investigating possible discrimination against girls. In contrast, it is argued that child mortality is a more reliable indicator for investigating gender discrimination. Section 5.2 discusses the proximate determinants of child mortality using cross-country data at the national level (Bhargava and Yu, 1997). Section 5.3 addresses conceptual issues in the analysis of household-level demographic data and presents findings for the proximate determinants of infant mortality using the NFHS-1 data from Uttar Pradesh, which is the most populous Indian state (Bhargava, 2003a). Issues of gender differentials in child mortality are also explored in the analysis. Section 5.4 addresses conceptual issues in modeling the proximate determinants of fertility, emphasizing the role of public and private health care providers (Bhargava, Chowdury *et al.*, 2005). The "endogenous facility placement" hypothesis is reappraised and it is argued that quality of health care infrastructure depends on the level of economic development in the regions. Empirical results are presented for the demand for contraceptives using a demographic (PERFORM) data set from India which compiled detailed information on various types of health care providers.

5.1 Identifying discrimination against girls in developing countries

In developing countries such as Bangladesh and India, researchers have reported a preferential treatment of boys (Chen *et al.*, 1981; Sen and Sengupta, 1983). Discrimination can take place in the allocation of food, medical treatment during illnesses, division of household tasks, etc. While this topic is of considerable policy importance, in practice it is complicated to investigate discrimination against girls. A popular approach for investigating discrimination is by comparing indicators of nutritional status for boys and girls. While one could also compare intakes of various nutrients, as discussed in Chapter 2, individuals' energy and nutrient requirements depend on their height, weight and levels of physical activity. Moreover, day-to-day variation in intakes can be high in developing countries due to poverty and food shortages (Bhargava, 1992). Thus, in investigating gender discrimination it is useful to focus on indicators of children's nutritional *status*, such as height and weight, and, more importantly, on child mortality.

Investigating gender discrimination using anthropometric variables

Measuring the percentages of children who are two (or more) standard deviations below the median compared to a reference population such as the US or the UK is a popular approach for identifying gender discrimination (e.g. Sen and Sengupta, 1983; Svedberg, 1990; Harriss, 1995). The z-scores for children's height and weight (equation (3.1)) can be computed using data on median heights and weights (and their standard deviations) from a reference population. While children's height and weight are good candidates for comparing the nutritional status of boys and girls in the same population, as discussed in Chapter 3, the quality of diet will differentially affect height and weight. It is, for example, possible for parents to favor boys by offering them more nutritious foods such as milk and meat, while girls may meet their energy needs primarily by consuming staple foods. Thus, z-scores, especially of height, may be useful for investigating possible gender discrimination.

There are certain methodological issues that need to be addressed before differences in the percentage of under-nourished boys and girls can be interpreted as substantive evidence of gender discrimination. For example, using US standards, investigators have found that a greater proportion of boys than girls were under-nourished in sub-Saharan African countries (Svedberg, 1990). This seemingly surprising result can be partially explained by noting that the dynamics of children's growth in the underlying reference population can influence the findings. For example, it is noted in the anthropometric assessment literature that children show different growth spurts at various ages (Tanner, 1986). In a cross-sectional setting, therefore, the numbers of boys and girls falling two standard deviations below the median will also depend on the dynamics of children's growth in the reference population. These issues are illustrated using z-scores based on US and UK standards applied to data on heights and weights from Vietnam and Pakistan. While one might *a priori* expect gender discrimination in Pakistan due to cultural factors, the dynamics of heights and weights in the reference population can contaminate the comparisons.

Results for children's heights and weights in Pakistan and Vietnam using the data from LSMS surveys

The data from the Living Standards Measurement Surveys (LSMS), conducted by the World Bank in Pakistan in 1991 and in Vietnam in 1993

Table 5.1 Percentages of children in Pakistan and Vietnam who are two standard deviations below median heights and weights in the US and UK

	HEIGHT							
	PAKISTAN				VIETNAM			
	US standards		UK standards		US standards		UK standards	
	Boy	Girl	Boy	Girl	Boy	Girl	Boy	Girl
All	46.6	45.7	50.7	47.4	54.7	52.8	56.4	57.3
Income 1[a]	51.7	49.3	56.5	52.2	62.6	61.4	65.0	66.4
Income 2	39.6	40.2	42.8	39.9	47.1	43.6	48.1	47.7
	WEIGHT							
	PAKISTAN				VIETNAM			
	US standards		UK standards		US standards		UK standards	
	Boy	Girl	Boy	Girl	Boy	Girl	Boy	Girl
All	50.1	45.3	53.0	55.4	51.7	47.3	57.3	61.5
Income 1[a]	52.9	47.2	57.2	56.5	57.5	53.9	63.0	67.5
Income 2	46.1	42.4	47.2	53.7	46.0	40.2	51.9	55.2

[a] Income group 1 consists of poorer households, and contains approximately half the children in the survey population.

Source: Previously unpublished results from author's research.

(World Bank, 1991, 1997), provide an opportunity to investigate differences by gender in children's heights and weights. The percentages of under-nourished children are calculated using the widely used US (National Center for Health Statistics) standards (Hamill *et al.*, 1977), and the standards for the UK (Freeman *et al.*, 1995). Table 5.1 presents the percentage of boys and girls who were two standard deviations below median heights and weights in the US and UK. Initially, the results are tabulated for all children in the sample. Subsequently, the sample is disaggregated into two income groups on the basis of per capita household incomes, with Income Group 1 consisting of poorer households.

The results for children's heights in Pakistan using US standards were close for boys and girls: roughly half the children fell two standard deviations below median US heights. This corroborates previous evidence in Alderman and Garcia (1993) using another data set from Pakistan. However, the use of UK standards suggested that a greater number of boys were under-nourished. In contrast with the results for Pakistan, a greater proportion of boys fell below the cutoff points when the US standards were used for the Vietnamese data. However, these differences were not striking, which may be partly due to the fact that coefficients of variation of heights are small. In absolute terms, Vietnamese children were shorter, which could be due to genetic differences. As expected, there were large differences between the proportions of under-nourished children in the two income groups in both countries.

The results using z-scores for children's weight are presented in the lower half of Table 5.1. For Pakistan, the US standards led to the conclusion that a higher proportion of boys were under-nourished. This ranking was reversed using the UK standards except for Income Group 1, where the results were close. Similarly, for Vietnam, the use of US standards showed that more boys were under-nourished, whereas one will conclude that more girls were below the cutoff points using UK standards. Overall, it is inadequate to compare the percentages of boys versus girls that are under-nourished using z-scores for heights and weights, due to the dynamics of the anthropometric measures in the underlying reference populations. Furthermore, one needs to compute confidence intervals for comparing differences in the z-scores of heights and weights. This is difficult in part because z-scores utilize different means and standard deviations for each age group in the reference population. Thus, variations in the estimated parameters in the reference population need to be taken into account in making the boy–girl comparisons. While one might simulate the confidence intervals, as a rough rule researchers should be concerned with gender discrimination if differences in the percentages of under-nourished boys and girls are quite large. Moreover, child mortality figures by gender can shed further light on gender discrimination.

Gender differences in child mortality in India

The above discussion has highlighted some difficulties in identifying discrimination against girls by using anthropometric data from Asian countries. An alternative approach is to focus on differentials in mortality rates for boys and girls. While it is natural that higher numbers of boys are born, boys are known to be less resilient to disease. Thus, relatively higher female mortality rates are likely to be indicative of possible discrimination against girls. The data from NFHS-1 conducted in 1992/3 in India can provide insights on this issue. The under-5 child mortality rates, expressed per 1,000 live births for boys and girls, were 115.4 and 122.4, respectively. The figures for the mortality of boys and girls in rural areas were 126.8 and 135.1; for urban areas, the corresponding figures were 77.5 and 79.3, respectively. These results suggest that girls may face some form of dis-

Table 5.2 Child mortality figures for the data from Uttar Pradesh using the NFHS-1 data (1982–1992)

	Total	Urban		Rural	
		Boys	Girls	Boys	Girls
Number of children born	21,620	2,059	1,898	9,133	8,530
Number of children died	3,165	167	185	1,343	1,470
	(14.6)	(8.1)	(9.7)	(14.7)	(17.2)
Age at time of death					
Less than 30 days	1,578	95	75	736	672
	(7.3)	(4.6)	(4.0)	(8.1)	(7.9)
1–6 months	459	24	31	200	204
	(2.1)	(1.2)	(1.6)	(2.2)	(2.4)
7–12 months	486	17	35	175	259
	(2.2)	(0.8)	(1.8)	(1.9)	(3.0)
13–24 months	379	14	27	137	201
	(1.8)	(0.7)	(1.4)	(1.5)	(2.4)
25–36 months	141	6	14	42	79
	(0.6)	(0.3)	(0.7)	(0.5)	(0.9)
37–48 months	81	7	3	34	37
	(0.4)	(0.3)	(0.1)	(0.4)	(0.4)
49–60 months	41	4	0	19	18
	(0.2)	(0.2)	(0.0)	(0.2)	(0.2)

Note: Figures in parentheses are the percentage of children who died.
Source: Bhargava (2003a).

crimination, especially in rural areas. A further breakdown of child mortality figures by age groups in the Indian state of Uttar Pradesh, which a high proportion are rural dwellers, would be informative.

Table 5.2 reports the number of children born in the period 1982–92 and the mortality figures in different age groups in Uttar Pradesh; the figures are presented separately for boys and girls in urban and rural areas (Bhargava, 2003a). Approximately 80% of children who died in the age group 0–5 years were less than 1 year old, i.e. infant mortality rates were high. Female child mortality in rural areas was higher in age groups 7–12 months and 13–24 months. Furthermore, the number of female deaths in rural areas in the age groups 13–24 months and 25–36 months was twice the corresponding figure for boys that had died. This was also true for urban areas, though the number of deaths was much smaller in absolute terms. Last, gender differentials in child mortality were negligible in the age groups 37–48 months and 48–60 months in both urban and rural areas. Overall, a significantly higher number of girls in rural areas died, especially in the age group 7–36 months. These figures probably reflect selective neglect of girls, possibly in the provision of health care when children are ill. However, to gain deeper insights into these issues, it is important to account for various confounding factors such as the socio-economic status of households, the birth order of the child, maternal and child vaccinations, etc. These issues are addressed in section 5.3, using the NFHS-1 data from Uttar Pradesh.

5.2 Longitudinal analyses of child mortality rates using national averages

The data at the national level on variables such as child mortality, fertility, literacy, gross domestic product (GDP) and expenditures on health care are widely available in databases of the United Nations and the World Bank (e.g. World Bank, 2005). While most variables are reported on an annual basis, demographic variables such as fertility and child mortality in developing countries are estimated from demographic surveys conducted in certain years. The figures reported for remaining years are projections based on assumptions on the trends in the variables. For example, Demographic and Health Surveys are carried out every five to ten years in developing countries and the in-between figures in the databases are interpolations. In fact, for many developing countries the United Nations (1992) presented data on child mortality rates based on demographic surveys in 1975, 1980 and 1985. A longitudinal analysis of these data at the national level can provide useful insights and guide further analyses using disaggregated data at regional and household levels.

From a conceptual viewpoint, specific hypotheses are often formulated in the biomedical field from casual observations of disease patterns. By contrast, economists are typically concerned with several related hypotheses that may be difficult to disentangle using aggregate data; macroeconomic analyses can provide useful insights for further investigations at the microeconomic level. Furthermore, policy-makers are interested in cross-country comparisons partly because the experiences of countries may be translated into useful polices for other countries without the need to carry out elaborate and expensive studies. However, cross-country analyses need to be carefully conducted since there are often large differences between the countries, and the results may be spurious, especially if the statistical issues have not been tackled. As a first step for identifying factors affecting child mortality in developing countries, it is useful to estimate longitudinal models using national averages, while controlling for unobserved between-country differences in the analyses. The empirical results can shed light on the relative importance of factors for reducing child mortality, such as female education, GDP and health care expenditures.

Results from a longitudinal analysis of national child mortality rates in developing countries

The results for infant and under-5 child mortality rates in African and non-African developing countries from estimating dynamic and static random-effects models using data in 1975, 1980 and 1985 are presented in Table 5.3 (Bhargava and Yu, 1997). It was necessary to estimate the models separately for non-African countries because data were unavailable on GDP (gross national product data were available for these countries). Moreover, there were likely to be large differences in the magnitudes of model parameters between African and non-African countries. Because the data on child mortality were available only at three time points, it was useful for assessing the robustness of the results to estimate static models where the lagged dependent variables were omitted from the set of explanatory variables. For dynamic models, the long-run elasticities of infant and child mortality rates with respect to explanatory variables are presented in Table 5.3. In contrast with the previous work by Aturupane *et al.* (1994), lagged dependent variables were treated as endogenous in the model. The use of appropriate estimation techniques circumvented

Table 5.3 Longitudinal regressions for infant and child mortality rates, African and non-African developing countries, 1975–1985

	African developing countries				Other developing countries			
	Dynamic random effects		Static random effects		Dynamic random effects		Static Random Effects	
Variable	q_1[a]	q_5[b]	q_1	q_5	q_1	q_5	q_1	q_5
Constant	2.573*	2.172*	1.287*	1.664*	0.420*	0.410	3.156*	3.197*
	(0.634)	(0.463)	(0.480)	(0.563)	(0.242)	(0.254)	(0.263)	(0.272)
Female illiteracy[c]	0.769	1.000	1.060	1.257	0.022	0.044	0.193	0.314
	(0.188)	(0.088)	(0.200)	(0.238)	(0.070)	(0.075)	(0.160)	(0.167)
Male illiteracy[d]	−0.139	−0.201	−0.169	−0.250	0.086	0.080	0.202	0.168
	(0.127)	(0.080)	(0.161)	(0.184)	(0.067)	(0.070)	(0.156)	(0.163)
RPCGNP[e]	−0.193	−0.142	−0.014	−0.014	−0.007	−0.005	−0.013	−0.010
	(0.067)	(0.051)	(0.025)	(0.023)	(0.007)	(0.007)	(0.009)	(0.009)
RGHELEXP[f]	−0.009	−0.076	−0.087	−0.166	−0.001	−0.002	0.018	0.018
	(0.033)	(0.026)	(0.027)	(0.031)	(0.014)	(0.015)	(0.028)	(0.030)
Time dummy 2[g]			−0.059	−0.109			−0.189	−0.240
			(0.038)	(0.033)			(0.040)	(0.042)
Time dummy 3[h]	−0.048	−0.036	−0.124	−0.168	0.063	0.106	−0.280	−0.328
	(0.017)	(0.026)	(0.055)	(0.049)	(0.042)	(0.050)	(0.043)	(0.045)
Lagged q[i]	0.110	0.123			0.802	0.800		
	(0.051)	(0.061)			(0.055)	(0.057)		
n	13	13	13	13	23	23	23	23
Long-run elasticities[j]								
Female illiteracy	0.864	1.140					0.193	0.314
RPCGNP	0.217	0.162					0.0	0.0
RGMELEXP		0.087	0.087				0.0	0.0

[a] Number of infants (per 1,000 live births) who died before reaching the age of 1 year.
[b] Number of children who died before age 5.
[c] Percentage of females over 15 years who were illiterate.
[d] Percentage of males over 15 who were illiterate.
[e] Real per capita GNP.
[f] Real government expenditure on health;
[g] dummy variable for 1980.
[h] Dummy variable for 1985.
[i] Lagged q_1 or q_5 included as a regressor in the dynamic model.
[j] Long-run elasticity = short-run elasticity/(1 − lagged q).
*$p < 0.05$.
Note: All variables were in natural logarithms and the reported coefficients are elasticities; standard errors are in parentheses. There were 3 repeated observations on 13 African countries (n = 13, t = 3) and 23 other developing countries (n = 23, t = 3).
Source: Bhargava and Yu (1997).

the surprising findings of Aturupane *et al.* (1994) that coefficients of lagged dependent variables were greater than 1.

The main findings in Table 5.3 were these. First, the elasticities of infant and under-5 child mortality with respect to female illiteracy were highly significant and close to 1 for African countries. By contrast, the corresponding elasticities for non-African countries were in the neighborhood of 0.20 and were significant in the static version of the model for under-5 child mortality. A possible explanation for the high magnitudes of elasticities in African countries was that many women in these regions are illiterate and have few opportunities to interact with educated women. By contrast, in non-African countries, interactions between illiterate and literate women can encourage illiterate women to adopt better hygiene, get vaccinated against tetanus during pregnancy and have their infants vaccinated. The male illiteracy rates, however, were not significant predictors of infant and child mortality rates in the models for African and non-African developing countries.

Second, real per capita GDP was a significant predictor of infant mortality rates in African countries in the dynamic model. By contrast, the GDP variable was not a significant predictor in non-African countries. It is known that infants born to under-nourished women often survive if basic health care is available. In African countries, the health care infrastructure is likely to improve significantly with increases in GDP and this may be reflected in significant coefficients of the GDP variable. Third, coefficients of real government expenditure on health were significant in the models for infant and under-5 mortality in African countries. However, this variable was insignificant in the models for non-African countries. A possible explanation may be that private health care services were generally available in non-African countries and these can help reduce child mortality. By contrast, most African countries are poor and households usually rely on the government to provide health care services. Fourth, the coefficients of lagged dependent variables were generally significant though the magnitudes were much smaller for African countries. The indicator variables for the years 1980 and 1985 were estimated with significant negative coefficients, indicating a decline in infant and child mortality rates over time in African and non-African developing countries.

In summary, the analysis of data at national levels supported the view that one is likely to see reductions in child mortality with increases in female education, though with some delays. However, analyses of cross-country data cannot afford modeling of the pathways through which female education can reduce child mortality. Moreover, between-household differences are obscured in such analyses. For example, well-off households in developing countries with access to health care services experience lower child mortality. By contrast, poor households not utilizing such services are likely to face higher child mortality. Analyses of cross-country data can identify factors, such as female education and income levels, that are important for reducing child mortality. However, the scope of such analyses is rather limited and it is important to conduct further studies using household-level data for understanding the determinants of child mortality. The next section presents an analysis of child mortality at the household level using the NFHS-1 data from Uttar Pradesh. The effects of female education, health care utilization and factors such as gender discrimination on child mortality are investigated in detail.

5.3 Proximate determinants of child mortality using the NFHS-1 household-level data

In section 5.1, the mortality rates for girls were seen to be higher than for boys, especially in rural regions of Uttar Pradesh. These differences may be due to factors that can be investigated using household-level data, such as from NFHS-1 in India. For example, girls born at low birth orders or "parity" (e.g. first or second child) may have lower mortality rates than those born at higher birth orders, especially if the latter children are "unwanted" (Bhargava, 2003a). Demographic surveys in developing countries enquire about couples' preferences for the "ideal" numbers of boys and girls (see below). Furthermore, composition of households in terms of surviving number of (older) boys and girls may differentially affect the survival chances of the "index" child (Muhuri and Preston, 1991). For example, older sisters can provide child care to their younger siblings, though this may not be the case for older brothers. Thus, it is important to develop a comprehensive framework for modeling the proximate determinants of child mortality using household-level data, and to investigate the effects of variables such as gender, household composition, socio-economic status and vaccinations on children's survival chances.

The National Family Health Survey-1 data from Uttar Pradesh

The NFHS-1 sample was representative of the Indian population and covered approximately 90,000 ever-married women in the age group 13–49 years from 25 Indian states. There were approximately 11,500 women from Uttar Pradesh, where roughly 80% of households lived in rural areas. There was retrospective information on households' background characteristics, landholding, caste, religion, dwelling space, possessions, etc. Information was gathered on certain sanitation and environmental variables such as the type of toilet used and the source for drinking water.

The data on child mortality were compiled using information on all live births during the previous ten years (1983–92). For every woman, information was gathered on education, age at marriage, fertility and family planning practices. Fertility preferences were investigated by posing hypothetical questions such as "How many of these children [the ideal number] would you like to be boys and how many would you like to be girls?" The answers to such questions were translated into indicator variables reflecting whether the child was "unwanted" (e.g. Bongaarts, 1990). For example, the indicator variable for whether a boy was "unwanted" was set equal to 1 if the boy was born at a parity where the preceding number of male births exceeded the "ideal" number of boys (Bhargava, 2003a); the indicator variable for "unwanted" girls was similarly defined. The women were also asked questions for investigating the desired family size, such as "Do you agree or disagree that an Indian family should have no more than two children?" Last, for the preceding five year period (1988–92), more detailed information was available on households' access to immunization programs and health care services. For example, variables such as whether the woman received ante-natal care, was visited by a health care worker during pregnancy, and was inoculated against tetanus were recorded. There was also information on whether the child was inoculated against polio, diphtheria and tetanus, place of delivery and complications during pregnancy were recorded.

Some analytical issues in modeling relationships between fertility and child mortality

The relationships between fertility and child mortality are dynamic and depend on the level of economic and social development in the region. Historically, when family planning methods and medical care were unavailable, maternal health and interactions between children's genotype and nutritional factors determined child survival. Thus, a large number of children were born and a relatively small proportion survived. The prospects of child survival have improved in recent years even in backward areas such as rural Uttar Pradesh. This will influence parents' perception of desired family size. For example, descriptive statistics from NFHS-1 showed that 65% of urban women and 56% in rural areas agreed that Indian families should not have more than two children. Moreover, 79% of urban women and 63% of rural women approved the use of family planning. The figures for actual use were 31% and 16%, respectively, presumably reflecting poor access to such services. Last, the ideal numbers of boys and girls in urban areas were 1.80 and 1.18, respectively. By contrast, these figures for rural areas were 2.22 and 1.35, respectively, indicating an especially greater preference for boys in rural areas.

In discussing the effects of child mortality on fertility (or vice versa), it is helpful to spell out the time frame in which the implications of the theory hold. For example, some proponents of the "child survival hypothesis" (Taylor *et al.*, 1976) have suggested that couples are unlikely to adopt family planning unless they are confident that their desired number of children will survive. This can be strictly true for irreversible procedures such as sterilization. More importantly, contraceptive use and its timing, while reflecting parental expectations regarding child survival, affect the survival chances of both existing and future children. For example, parents in affluent societies can be confident of child survival shortly after birth. By contrast, child mortality is high in developing countries and the survival pattern is more complex. The under-5 child mortality figures in Table 5.2 show that 91% of children in Uttar Pradesh who died were less than 2 years old. Thus, it can take two to three years for survival uncertainties to be resolved; irreversible family planning procedures are unlikely to be attractive in developing countries unless couples have surpassed their reproductive goals.

Furthermore, if family planning services are unavailable in the period following a birth, then risk of pregnancy is high. The ensuing pregnancy will demand replenishment of vital nutrients such as iron and calcium to support fetal growth (Scrimshaw, 1996). This is difficult in many developing countries because the bioavailability of nutrients such as iron from staple foods is low (Bhargava, Bouis *et al.*, 2001); nutrient losses due to infections can also diminish women's capacity to produce healthy infants. Depending on parity, frequent pregnancies can reduce the time available for child care and subsistence activities. Thus, unmet need for family planning is likely to result in a large number of children born at short time intervals; short birth intervals are associated with increased risk of child mortality in the demographic literature (e.g. Hobcraft *et al.*, 1983; Gribble, 1993). If, however, women use family planning, then births can be spaced. This will be beneficial for the health of the surviving children and improve the prospects of carrying subsequent pregnancies to full term.

At a given point in time, the direction of causality in the fertility–mortality relationship has asymmetric implications for infant mortality. High mortality at low-order births is likely to influence the parents' decision to have more than the desired number of children. By contrast, if mortality is the result of a large number of unwanted births, then children born at higher parities will be at greater risk. A corollary of these observations is that when the historical data on child mortality and fertility are analyzed, decline in child mortality rates is likely to precede decline in fertility. By contrast, when data from household surveys are analyzed, high fertility, especially among poor households with limited access to health care services, is likely to exacerbate child mortality. Moreover, in analyses of household-level data, there may be gender differences in child mortality; female mortality at higher parities in Punjab was high (Das Gupta, 1987). Also, mortality rates for first-born children can be high (Bongaarts, 1987; Trussell, 1988); from a biological viewpoint, there is competition for nutrients between the fetus and the young mother's own requirements for growth (Falkner *et al.*, 1994). Multivariate analyses can partially control for this by including maternal age at first birth as an explanatory variable in the model.

Last, in the absence of access to family planning, the total number of live births to a woman will depend on her fecundity, nutritional status, age at marriage and breast-feeding patterns. The number of surviving children is affected by demographic, socio-economic and health care variables. Another important aspect of modeling the proximate determinants of infant survival is that parents may desire a certain number of sons (Chen *et al.*, 1981; Sen and Sengupta, 1983). This led Muhuri and Preston (1991) to postulate that the number of surviving older brothers and sisters will differentially affect the survival chances of the index child. It is common in poor households for young girls to spend time on household tasks. This may not be the case for boys, especially in rural areas where adult males undertake strenuous work. Because intra-uterine growth retardation is likely to increase with parity (Al *et al.*, 1997; Chapter 4), mothers may have less time to care for children born at high parities. Thus, older sisters can fill the child care gap. While the survival chances of siblings are likely to be positively correlated, care given by girls will strengthen the coefficient of older sisters in a model explaining the index child's survival chances. Moreover, the numbers of surviving older brothers and sisters may be correlated with the error terms affecting the models. These issues can be addressed using econometric methods discussed briefly in the next subsection, which can be skipped by non-technical readers.

Sources of endogeneity and econometric estimation

A possible source of simultaneity in the infant survival relationship is that determinants of the survival of older siblings and the index child are likely to be similar. Also, the number of surviving children before a family planning method is used is potentially an endogenous variable, since it reflects parental preferences about family size. While it is important to address endogeneity issues, the results will also be presented from models with dichotomous dependent variables that ignore the endogeneity of explanatory variables; the non-technical reader can focus on these results since they are close to the results from more complex models. Moreover, a satisfactory treatment of endogenous explanatory variables entails the use of "instrumental variables" that are highly correlated with endogenous variables. This issue is important in probabilistic models and the focus of this subsection is on substantive factors affecting infant survival.

A probit model explaining the survival to age (say) 1 year of N live births (y_i) by m endogenous variables (Y_i) and n exogenous variables (X_{1i}) is given by equations (5.1) and (5.2):

$$y_i = Y_i'\gamma + X_{1i}'\beta + u_i \quad (i = 1,2,\ldots,N) \tag{5.1}$$

$$Y_i = P'X_i + V_i \quad (i = 1,2,\ldots,N) \tag{5.2}$$

Equation (5.1) is the "structural" model for the dichotomous variable y, which is unity if the child dies before reaching the age of 1, and zero otherwise. The coefficients of Y and X_1 are represented by $m \times 1$ and $n \times 1$ vectors γ and β, respectively. Equation (5.2) is a "reduced form" for m endogenous variables in equation (5.1), with the $m \times n$ coefficient matrix P'; V is an $m \times 1$ vector of reduced-form errors. It is assumed that certain exogenous variables are excluded from the structural model for identifying parameters (Newey, 1987; Rivers and Vuong, 1988). The errors on equations (5.1) and (5.2) are assumed to be jointly normally distributed.

Furthermore, assuming that there are H households in the sample with a different number of live births, the errors affecting equations (5.1) and (5.2) will have an "unbalanced" structure, i.e. a different number of children born in different households. Treating household effects as randomly distributed variables, the errors on equation (1) can be decomposed as

$$u_{hj} = \delta_h + w_{hj} \quad (j = 1,\ldots,J_h; \ h = 1,2,\ldots,H) \tag{5.3}$$

where δ's are household-specific random effects, w's are randomly distributed variables, and J_h is the number of live births in household h. Similarly, random effects can be introduced in reduced-form equations (5.2); errors on the first endogenous variable in (5.2) can be written as

$$v_{1hj} = \lambda_{1h} + w_{1hj} \quad (j = 1,\ldots,J_h; \ h = 1,2,\ldots,H) \tag{5.4}$$

In order to tackle the endogeneity of explanatory variables included in equation (5.1), researchers have derived the "conditional" likelihood function by assuming that the errors on equations (5.1) and (5.2) are jointly normally distributed (Smith and Blundell, 1986). This entails working with the density of u_i conditional on V_i in equation (5.1), and substituting a consistent estimator of P from equation (5.2) in the modified equation (5.1). The model parameters are estimated by maximum likelihood using a numerical optimization scheme; standard errors are obtained using a Taylor series expansion. The conditional likelihood function with random effects was developed by Vella and Verbeek (1999); software packages (e.g. LIMDEP, 1995; STATA, 2003) can be used to estimate the model parameters. Because the errors have a random-effects structure, it is necessary to include two covariance terms per endogenous variable in equation (5.1). The exogeneity of a variable is tested by testing the null hypothesis that coefficients of covariance terms computed from the reduced form residuals are zero.

Empirical results from models for infant mortality in Uttar Pradesh in the previous five years

The empirical results from estimating probit models for infant mortality in the previous five years (1988–92) are presented in Table 5.4; the results from models addressing endogeneity problems were similar, so that one can focus on the substantive findings. First, the number of surviving older brothers and sisters was positively associated with the survival chances of the index child. Since older sisters are likely to take care of younger siblings, one would expect the coefficient of girls to be higher than the corresponding estimate for boys, which was the case. The null hypothesis that the coefficients of boys and girls were the same was rejected by a likelihood ratio test. There were non-linearities in the relationships between infant survival and the number of surviving older sisters. For example, the squared term for older sisters was estimated with a negative coefficient that was significant at the 10% level, indicating an adverse effect after couples have had a certain number of girls.

Second, birth interval was significantly positively associated with infant survival. The number of surviving children before a family planning method was used was negatively and significantly associated. If the woman did not use family planning, this variable was set equal to the number of surviving children. Because the model accounted for the survival status of children

Table 5.4 Maximum-likelihood estimates of probit models for infant survival in Uttar Pradesh (1988–1992)

Variable	Model 1	Model 2	Model 3[a]	Marginal effects[b]
Constant	0.912*	0.912*	1.103*	
	(0.165)	(0.164)	(0.167)	
Indicator for rural areas	−0.037	−0.037	−0.047	−0.005
	(0.075)	(0.075)	(0.077)	(0.008)
Indicator for girls	0.114*	0.114*	0.170*	0.018*
	(0.063)	(0.063)	(0.068)	(0.007)
Indicator for girls born	−0.208*	−0.208*	−0.261*	−0.028*
after the ideal number	(0.066)	(0.065)	(0.070)	(0.008)
Indicator for boys born	−0.016	−0.016	−0.035	−0.004
after the ideal number	(0.059)	(0.060)	(0.061)	(0.007)
Number of older girls	0.185*	0.184*	0.227*	0.025*
	(0.024)	(0.022)	(0.030)	(0.004)
Number of older boys	0.125*	0.126*	0.162*	0.018*
	(0.024)	(0.023)	(0.032)	(0.004)
Mother's education	0.0001	0.0002	−0.001	−0.0001
	(0.020)	(0.020)	(0.020)	(0.002)
Mother's age at first birth	0.002	0.002	−0.001	−0.001
	(0.007)	(0.007)	(0.007)	(0.001)
Birth interval	0.180*	0.180*	0.186*	0.018*
	(0.016)	(0.014)	(0.014)	(0.002)
Number of children before	−0.124*	−0.124*	−0.138*	−0.015*
family planning	(0.014)	(0.012)	(0.017)	(0.002)
Indicator for tetanus vaccination	0.289*	0.289*	0.282*	0.031*
	(0.047)	(0.047)	(0.049)	(0.005)
Indicator for child vaccination	1.001*	1.001*	0.988*	0.107*
	(0.063)	(0.066)	(0.069)	(0.007)
Indicator for no toilet	−0.117	−0.117	−0.104	−0.011
	(0.072)	(0.070)	(0.072)	(0.008)
Number of rooms	0.007	0.007	0.006	0.001
	(0.009)	(0.009)	(0.009)	(0.001)
Covariance term 1 (girls)			0.053*	0.006*
			(0.014)	(0.001)
Covariance term 2 (girls)			−0.280*	−0.030*
			(0.018)	(0.002)
Covariance term 1 (boys)			0.223*	0.024*
			(0.024)	(0.003)
Covariance term 2 (boys)			−0.094*	−0.010*
			(0.014)	(0.002)
Rho[c]		0.001	0.127	
		(22.817)	(0.182)	

Note: There were 9,674 live births.
[a] Model 3 treats number of older boys and girls as endogenous.
[b] This column reports the "marginal effects" for parameter estimates in Model 3. Asymptotic standard errors are in parentheses; covariance terms are conditional expectations of errors in equation (5.1).
[c] The proportion of variance due to household-specific random effects.
*$P < 0.05$.
Source: Bhargava (2003a).

born before the index child, early use of family planning was beneficial for infant survival. The estimated coefficients of indicator variables for whether the mothers were vaccinated against tetanus and whether the children were vaccinated were large and statistically significant; thus, health care programs seem critical for infant survival. The analysis was repeated using data for the period 1988–91 (excluding births in 1992) to avoid the problem that some infants may have died shortly after the survey. However, the empirical results for the past four and five years were very close. Because children were typically vaccinated around the age of 3 months, the indicator for child vaccination was dropped from the models to investigate the robustness of the results. This led to a change in the coefficient of the indicator variable for maternal tetanus vaccination, though the remaining estimates were similar.

Fourth, the coefficients of maternal education and age at first birth, the indicator variable for not having a toilet, and the number of rooms in the house were not statistically significant in Table 5.4, though these coefficients were significant when the data from the previous ten years were used (Bhargava, 2003a). The lack of statistical significance could be due to the reduction in the sample size. Also, more educated women were likely to utilize health care services, so that the effects of female education on infant survival were likely to operate via greater utilization of such services. Last, while the indicator variable for rural areas was insignificant in Table 5.4, the indicator variable for girls was significant, though with a *positive* sign. By contrast, the indicator variable for girls born at a higher parity than the ideal number had a *negative* coefficient. These results suggested that girls born at low parities were not at a disadvantage, which could be due to better maternal nutritional status and also because daughters' contribution to housework may be viewed favorably. This would not be the case for girls born at high parities, who face greater growth retardation and increased competition for resources, including medical care. The indicator variable for boys born after the ideal number was not significant in Table 5.4, suggesting a selective neglect of girls born at high parities. It is evident that econometric modeling of demographic data can facilitate a better understanding of factors underlying gender differentials in child mortality in India, where there may be a preference for boys, especially in rural areas.

In summary, the results presented in Chapter 5.3 pointed towards two main conclusions. First, the indicator variables constructed using the stated preferences for sons and daughters indicated that unwanted fertility contributed to excess mortality of girls born at higher birth orders in Uttar Pradesh. Family planning programs are therefore likely to narrow gender differentials in infant mortality, especially if sex-selective abortions are not available. Such conclusions could not have been drawn using the data on anthropometric indicators (section 5.1) or from the descriptive statistics on child mortality (Table 5.2). Second, while maternal education was not a significant predictor of infant survival in some of the models, immunization programs had a strong beneficial impact. Because a majority of women in rural Uttar Pradesh were uneducated, adult education programs would require substantial resources. Investments in family planning and immunization programs are expensive but are likely to prevent infant mortality in a shorter time frame. Of course, female education should be an important policy goal for improving children's health status and cognitive development (Chapter 3).

5.4 Health care infrastructure and fertility

The results from estimating models for infant mortality in Uttar Pradesh indicated the importance of variables, such as birth intervals and vaccinations, that depend on women's access to health care services. While most demographic surveys compile some information on the health care infrastructure, detailed information on the quality of services available in public and private clinics is seldom available. Moreover, in developing countries numerous private "agents", such as medical shops and independent practitioners, provide health care. Thus, it is complicated though important to assess the effects of overall health care infrastructure on utilization of health care services.

From a conceptual viewpoint, following the work by Becker (1982), economists have emphasized that family planning programs in developing countries may be "endogenously" placed in response to health conditions in the region (e.g. Angeles *et al.*, 1998). For example, if public clinics are placed in areas with high child mortality, then the errors affecting a model for child mortality may be correlated with households' distances from the facility. While governments often require placement of public health facilities on the basis of population (Koenig *et al.*, 2000), the quality of services is likely to depend on economic development in the region. It is often difficult to place skilled medical personnel in remote areas; the quality of services in such areas is likely to be poor, while child mortality is likely to be high. Furthermore, in developing countries such as India, there has been a rise in private health care providers (Peters *et al.*, 2002), who are driven by profit motives and can influence the quality of services in public facilities. Such issues merit analytical and empirical investigation; the poor quality of public services in backward regions and the increased role of private providers have different implications for child mortality than models emphasizing endogenous placement of facilities. In the next subsection, certain conceptual issues are discussed and results are summarized from analyses of the PERFORM data set from Uttar Pradesh, which contains extensive information on health care providers (Bhargava, Chowdury *et al.*, 2005).

Demand and supply schedules for contraceptives

Easterlin and Crimmins (1985) underscored the role of economic factors which affect the demand and supply schedules of children. With economic development, there is clearly a need to educate children for gaining employment in skilled occupations. This, together with a decline in child mortality and access to health care services, is likely to lower desired family size. Moreover, cultural factors play an important role in the adoption of family planning methods (Caldwell, 1982; Cleland and Wilson, 1987). However, since couples are postulated by Easterlin and Crimmins (1985) *both* to demand and supply their children, it is important to re-appraise some of the postulates in their analyses of fertility behavior.

The supply schedule is used in economics to deduce quantities of commodities offered at different prices, holding other relevant variables constant. However, the number of children born to a married woman mainly depends on her fecundity, which in turn is influenced by biological processes that are not amenable to empirical modeling. By contrast, the desired number of children is likely to depend on economic factors, such as expenditures that households can afford on food, clothing, education and health care. Furthermore, the *actual* number of children born to a woman is influenced by the efficacy of the contraceptives used. Because factors affecting the supply schedule for children are opaque, and demand for contraception is a "derived" demand for avoiding unwanted births, it is more appealing to analyze fertility behavior using demand and supply schedules for contraceptives rather than for children.

Further, there are differences between demand for commodities that yield utility or satisfaction and the demand for contraceptives that are inconvenient to use. It is perhaps not surprising that most couples in backward states such as Uttar Pradesh rely on female sterilization after surpassing their fertility goals. This, however, does not diminish the importance of contraceptives for birth spacing. First, it takes approximately two years after a birth in Uttar Pradesh to ensure that the child will survive (Table 5.2). If contraceptives for birth spacing are unavailable, then the woman could get pregnant, thereby increasing her unwanted fertility. Second, demand for contraceptives for birth spacing and that for terminal methods merit separate treatments because the supply of these services depends on different factors. Last, terminal methods for family planning, such as sterilization, require skilled services and entail risks such as sexual dysfunction. The availability of qualified staff and adequate drugs are essential for encouraging couples to opt for such procedures; fees for services in private clinics can hinder utilization. Thus, variables reflecting service quality in public and private facilities are likely to affect the use and timing of terminal methods. By contrast, the demand schedules for birth control pills and condoms are influenced by economic factors and couples' awareness of benefits from birth spacing.

Gradual evolution of the health care infrastructure versus endogenous facility placement

In most developing countries, the health care infrastructure has evolved gradually over time and consists of public facilities, private providers and non-governmental organizations (NGOs). Initially, health care is available mainly in urban areas via government facilities and private practitioners offering services to those who can afford them. Because life in urban areas affords opportunities such as education for children, urban areas are attractive venues for medical personnel. A concentration of public and private facilities staffed by qualified personnel is likely to increase competition among providers, thereby enhancing the quality of services. Even poor people seeking treatment for serious illnesses may spend their savings in private facilities if public care is viewed to be inferior.

In the absence of major public investments, the health care infrastructure in rural areas is likely to evolve very slowly. While governments mandate the placement of hospitals or community health centers based on population, quality of services is likely to depend on the allocated resources and the willingness of qualified personnel to serve in remote areas. The low purchasing power of households in underdeveloped areas reduces the incentives for private providers to set up facilities. To fill the health care gaps, NGOs supported by the government and other agencies often deliver basic services. The quality of services, as measured, for example, by the number of trained personnel providing health care, is likely to be poor. This will be reflected in lower utilization rates and higher child and maternal mortality.

In contrast with the gradual evolution of the health care infrastructure, the literature on endogenous facility placement emphasizes the role of pressure groups for the placement of public facilities. Such formulations, however, seem applicable only to situations where the government is the sole provider of health care and acts appropriately without delays. Moreover, governmental efforts to improve service quality are likely to be hampered by logistical difficulties in transferring equipment to remote places, which may lack even electricity. The facts that a high proportion of health care may be privately provided and that quality differences between services have been unexplored suggest that it would be useful to analyze the effects of health care infrastructure on contraceptive use and infant mortality in a broader analytical framework. The proximate determinants of fertility are analyzed in the next subsection using the PERFORM data from Uttar Pradesh, which contain extensive information on public and private health care providers.

The PERFORM data from Uttar Pradesh

The PERFORM survey was designed to measure indicators of reproductive health at three levels. First, the data were compiled on public and private providers of health care. Second, the work experience of staff in the delivery points was investigated. Finally, detailed information was compiled on married women who were likely to utilize health care and family planning services. The survey was conducted in 1995 in 1,539 villages and 738 urban blocks within 1,911 Primary Sampling Units (PSUs), interviewing 40,633 households, 2,428 fixed service delivery points and 6,320 staff members, and 22,335 individual service agents such as health workers and medical shops (SIFPSA, 1996). The urban and village questionnaires investigated the number of households, clinics, private practitioners, cooperatives, voluntary organizations and industries in the region. The household questionnaire investigated the demographic composition of the household, landholding and other variables. The women's questionnaire covered variables such as marital status, reproductive history, access to health care and family planning services, quality of services, fertility preferences and contraceptive use. There was detailed information on up to three births in the previous three-year period. Vaccinations against tetanus and complications during and after pregnancy were recorded.

The Fixed Service Delivery Point questionnaire investigated the availability of services such as male and female sterilization, IUD insertion and medical termination of pregnancy. Providers were mapped to households in the PSU. The number of months of supply of contraceptives, such as condoms and birth control pills, in public and private facilities and by private agents were calculated from the responses. Staff in the facilities, including doctors, nurses and social workers, were interviewed to assess their qualifications. Several indices were constructed to approximate the health care infrastructure. The number of allopathic doctors (trained in "Western" medicine) performing male or female sterilization, inserting IUDs and terminating pregnancies was calculated separately for government hospitals, community health centers, private hospitals and private agents. The total numbers of staff as well as the averages calculated over the respective types of facilities were used as indicators of the health care infrastructure; the use of averages enabled comparisons across groups. Similarly, the numbers of 'full-time equivalent' staff and those devoted to family planning services in public and private facilities were calculated. (Bhargava, Chowdhury *et al.*, 2005).

The sample means and standard deviations of selected variables are displayed in Table 5.5. Approximately 68% of the women had never attended school; the average number of children was 3.49 in the subsample covering contraceptive use. Female sterilization was the most common method of family planning, with 15% of women opting for the procedure; only 1.4% of men were sterilized. The percentage of couples using IUD, birth control pills and condoms was 1.5, 1.7 and 3.7, respectively. The average number of government and private hospitals for the PSU were 0.63 and 0.13, respectively. While there were several private agents operating in the PSU, on the basis of qualifications to perform sterilization, terminate pregnancies and insert IUDs, the average for the PSU was 0.029. The average number of family planning staff available in government hospitals, community health centers and private hospitals was 2.45, 11.03 and 1.72, respectively. The average months' supply of birth control pills and condoms was approximately 0.21 for government and private hospitals and private agents, and 0.54 for community health centers.

Models for fertility and the quality of care in public and private sectors

The proximate determinants of the use of family planning methods were analyzed via binary regressions (Bhargava, Chowdhury *et al.*, 2005). The empirical model for the chances of female sterilization (or IUD use) is given by:

TABLE 5.5 Sample means (or percentages) and standard deviations of selected variables in the models for contraceptive use and infant mortality estimated from the PERFORM survey from Uttar Pradesh

	Mean or %	SD
Woman's age in years	29.74	8.54
Ever attended school (1 = Yes, 0 = No)	0.32	0.47
Number of children born	3.49	2.27
Number of surviving children	2.99	1.85
Household possessions index (0–6)	2.67	1.79
Backward caste (1 = Yes, 0 = No)	0.51	
Woman sterilized (1 = Yes, 0 = No) (%)	15	
Man sterilized (1 = Yes, 0 = No) (%)	1.4	
Woman using IUD (1 = Yes, 0 = No) (%)	1.5	
Woman using birth control pills (1 = Yes, 0 = No) (%)	1.7	
Man using condoms (1 = Yes, 0 = No) (%)	3.7	
Number of women with a birth in the 3-year period	19,620	
Children who died before reaching age of 1 year (%)	3.9	
Births wanted "later" (%)	7.5	
Births wanted "never" (%)	5.5	
Birth interval (years)	3.09	2.07
Woman vaccinated against tetanus (%)	54	
Number of women with 2 births in the 3-year period	2,037	
Second birth wanted "never" (%)	5.9	
Average number of government hospitals in PSU	0.63	2.58
Average number of private hospitals in PSU	0.13	0.58
Average number of private allopathic doctors in PSU	0.33	0.85
Average number of private doctors trained in terminal FP methods in PSU	0.029	0.024
Average family planning staff: government hospital	2.45	7.50
Average family planning staff: private hospital	1.72	7.20
Average family planning staff: community health center	11.03	60.37
Months supply of birth control pills and condoms: government hospitals	0.21	1.08
Months supply of birth control pills and condoms: private hospitals	0.21	1.09
Months supply of birth control pills and condoms: community health centers	0.54	1.52
Number of private agents trained in dispensing birth control pills in PSU	0.22	0.82

Note: There were 30,966 and 19,632 women in the samples for contraceptive use and infant mortality, respectively. PSU = Primary Sampling Unit.

Source: Bhargava, Chowdhury et al. (2005).

$$(\text{Female sterilization})_i = a_0 + a_1(\text{Woman's age})_i + a_2(\text{Ever attended school})_i$$
$$+ a_3(\text{No. of surviving children})_i$$
$$+ a_4(\text{No. of surviving children})_i^2$$
$$+ a_5(\text{Possessions index})_i$$
$$+ a_6(\text{Backward caste})_i$$
$$+ a_7(\text{No. of govt hospitals})_i \qquad (5.5)$$
$$+ a_8(\text{No. of pvt hospitals})_i$$
$$+ a_9(\text{Avg. qualified staff: govt})_i$$
$$+ a_{10}(\text{Avg. qualified staff: pvt})_i$$
$$+ a_{11}(\text{Avg. qualified staff: pvt agent})_i$$
$$+ u_{1i}(i = 1, \ldots, N)$$

113

The variable "Ever attended school" was an indicator (0/1) variable that was 1 if the woman had attended school; "Backward caste" was an indicator variable that was 1 if the household belonged to a scheduled caste or a tribe; an indicator variable for rural/urban location of the household was not significant in these models. The staff qualified to perform sterilization, terminate pregnancy and insert IUDs were calculated for each facility; "Average qualified staff" variables in equation (5.5) were averages over the respective facilities, and for private agents in the PSU. The error terms u_{1i} were assumed to be distributed as a logistic distribution for the N women in the sample; u_{1i} were assumed to be normal for probit models. Similarly, models were developed for the uses of pills and condoms. The models for IUD, birth control pill and condom use were re-estimated, dropping sterilized women from the sample. Also, ordinal and multi-nomial regressions models were estimated for addressing the issues of choice between various family planning methods.

Empirical results for contraceptive use from the PERFORM data set from Uttar Pradesh

The maximum-likelihood estimates from binary logistic models for the chances of women opting for sterilization, and for the use of IUD, birth control pills and condoms, are presented in Table 5.6. In the model for female sterilization, the woman's age and having attended school significantly increased (P <0.05) the chances of sterilization. There were non-linearities with respect to number of surviving children in that the birth of an additional child increased the chances of sterilization, though at a decreasing rate. The household possessions index was estimated with a positive and statistically significant coefficient. While the average number of government hospitals for the PSU was not a significant predictor, the average number of private hospitals was positively and significantly associated with the chances of women opting for sterilization. The average number of private doctors trained in performing sterilization, IUD insertion and terminating pregnancies significantly increased the chances of female sterilization. By contrast, the availability of such personnel in government and private hospitals were not significant predictors of female sterilization. Because the indices based on PERFORM data on health care infrastructure primarily reflected the availability of services in the PSU, these results highlighted the importance of access to trained medical

TABLE 5.6 Maximum-likelihood estimates of binary logistic regressions for the use of female sterilization, IUD, birth control pills and condoms, explained by demographic, socio-economic and health care infrastructural variables using the PERFORM survey

	Female sterilization		IUD		Birth control pills		Condoms	
	Coefficient	SE	Coefficient	SE	Coefficient	SE	Coefficient	SE
Constant	−7.043*	0.115	−4.437*	0.239	−3.026*	0.199	−3.248*	0.149
Woman's age	0.082*	0.002	−0.036*	0.008	−0.063*	0.007	−0.040*	0.005
Backward caste	0.010	0.034	−0.664*	0.170	−0.380*	0.091	−0.398*	0.069
Ever attended school	0.430*	0.040	0.835*	0.108	0.505*	0.097	0.930*	0.073
Number of children surviving	1.300*	0.046	0.318*	0.104	0.168*	0.029	0.045*	0.023
(Number of children surviving)^2	−0.150*	0.005	−0.040*	0.014				
Household possessions index	0.121*	0.010	0.254*	0.030	0.136*	0.027	0.245*	0.021
Average number of government hospitals	0.014	0.076	−0.557*	0.283	−0.434	0.267	−0.116	0.156
Average number of private hospitals	0.088*	0.025	0.059	0.064	−0.027	0.071	0.112*	0.041
Average number of private doctors trained in terminal FP methods	0.053*	0.024	0.080*	0.036				
Average number trained in FP methods: government hospitals	0.038	0.025	0.200*	0.049				
Average number trained in FP methods: private hospitals	−0.033	0.022	−0.064	0.052				
Months supply pills (condoms): government hospitals					−0.052	0.134	0.008	0.010
Months supply pills (condoms): community health centers					−0.024	0.030	−0.055*	0.025
Months supply pills (condoms): private hospitals					0.109	0.094	0.156*	0.023
Number of private agents trained in pills and condoms					0.188*	0.083	0.073	0.065
R^2	0.198*		0.102*		0.043*		0.118*	

Note: There were 30,966 women in this sample; slope coefficients and standard errors are reported.
*P <0.05.

Source: Bhargava, Chowdhury et al. (2005).

personnel in both public and private sectors for increasing the chances of female sterilization.

The results for IUD use were similar in many respects to those for female sterilization but there were some differences. For example, the coefficient of woman's age was estimated with a negative sign, indicating that younger women were more likely to use IUD, presumably for birth spacing. Also, women from backward castes and tribes had significantly lower chances of IUD use. Coefficients of the indicator variable for school attendance and number of surviving children were similar in the models for female sterilization and IUD use. The average number of trained private doctors and of staff available in government hospitals were positively and significantly associated with IUD use.

The results for the use of birth control pills and condoms were similar. The relationships were linear in the number of surviving children; older couples were significantly less likely to use birth control pills and condoms. The household possessions index was positively and significantly associated with birth control pill and condom uses. However, women from backward castes and tribes were less likely to use these methods. While the supplies of pills and condoms in government hospitals and community health centers were not significantly associated with pill use, the number of trained private agents was a positive and significant predictor. The number of months of supply of condoms in private hospitals was positively and significantly associated with condom use. This was not true for the supply of condoms in government hospitals and in community health centers. Because birth control pills and condoms can be obtained in a more discreet manner from private suppliers by households that can afford them, it was perhaps not surprising that the results underscored the private sources. The (pseudo) R^2 (Cox and Snell, 1989) for the model for condom use was higher than that for birth control pill use, partly because a greater number of couples were using condoms. Last, dropping sterilized women from the sample led to very similar results in the models for IUD, birth control pill and condom use. Also, the use of ordinal and multinomial regressions led to similar findings (Bhargava, Chowdhury *et al.*, 2005).

5.5 Conclusions

This chapter covered several aspects of fertility and child mortality relationships in developing countries, with emphasis on possible discrimination against girls in India. The role of health care infrastructure for lowering fertility and child mortality rates was underscored. While it is true that educated women are likely to desire fewer children and their children are at lower risk of mortality, it is essential that governments in developing countries invest in the health care infrastructure to increase the uptake of services. This would be beneficial for the health of surviving children, which in turn is critical for their cognitive development and the future supply of skilled labor (Chapter 4). Such issues are not sufficiently emphasized in the economics and demographic literature. For example, as seen in Chapter 4, nutritional interventions, such as providing fortified foods to schoolchildren in Guatemala, brought lasting benefits. However, the long-run efficacy of such programs would be enhanced if mothers used family planning to space births, which, in turn, would lead to healthier infants and also mothers having more time for child care. Moreover, nutritional interventions are inevitably done on a small scale, while a large number of inhabitants of developing countries require health care and family planning services. Thus, demographic factors play a critical role in economic development, especially in countries with high fertility rates. This is especially important for certain Indian states such as Uttar Pradesh and for sub-Saharan African countries such as Ethiopia, which are endowed with few natural and human resources (Bhargava, 2007). Finally, on the basis of the results from cross-country analysis in section 5.2 and the results in sections 5.3 and 5.4, one should not view analyses of cross-country data on *average* fertility rates in developing countries as casting doubt on the efficacy of family planning programs (Pritchett, 1994). As seen in this chapter, fertility depends on socio-economic variables and on access to family planning services, and such factors are masked when cross-country data on fertility rates are analyzed.

6

Nutrition, health and productivity in developing countries

The effects of the nutritional and health status of adults on labor productivity in developing countries are of utmost importance from a policy standpoint. If, for example, adults are severely under-nourished and cannot earn an adequate living, then all household members, including children, will be at severe risk of malnutrition. Moreover, under-nourished adults have reduced immunity to infections (Scrimshaw *et al.*, 1959; SanGiovanni and Scrimshaw, 1997), and may not be able to perform strenuous tasks for long periods. Even for individuals in less strenuous occupations, nutrient deficiencies and poor environmental conditions, such as lack of sanitation, can increase sicknesses, thereby leading to productivity loss. From a historical perspective, Fogel (1994) attributes large gains in economic productivity to improvements in the nutritional status of populations. In a similar vein, Floud *et al.* (1991) found a significant impact of sicknesses on health indicators, such as height in the UK, over the last three centuries. Thus, one would expect to see higher productivity among well-nourished populations, especially in developing countries, where there are often large variations in indicators of nutritional and health status such as anthropometric measures, dietary intakes and morbidity.

It was seen in Chapter 2 that increases in household incomes lead to improvements in diet quality in that intakes of protein and micronutrients increased with incomes in India, the Philippines and Kenya. This chapter is concerned with the converse of this phenomenon, i.e. productivity (or income) gains resulting from higher energy and nutrient intakes. The analytical issues are complex for several reasons. First, one needs to distinguish between the effects of energy and nutrient intakes in short versus long time frames. For example, even if energy intakes are adequate to perform certain tasks, if the intakes of critical nutrients such as protein, iron and vitamins A and C have been consistently below required levels, then individuals' long-term health status may be compromised. Such individuals may not be able to undertake strenuous tasks even if their energy intakes are increased. This is a common problem in developing countries, where many adults stop performing strenuous tasks from a relatively early age. Second, the observed balance between energy intakes and expenditures is complicated by the fact that even under-nourished individuals carry around 5 kg of body fat (i.e. 45,000 kcal of stored energy) so that energy intakes and expenditures need not be equal at all points in time. For example, higher energy intakes on a particular day need not imply higher energy expenditures and vice versa. Thus, it is essential to interpret the effects of nutritional status on labor productivity in a broad analytical framework. Section 6.1 begins by discussing certain studies conducted by biomedical researchers on the effects of nutritional status on labor productivity in developing countries. The approaches used in economics are briefly discussed in section 6.2. Last, the effects of adults' nutritional status on time allocation patterns are discussed in section 6.3, using an elaborate longitudinal data set from Rwanda (Bhargava, 1997).

6.1 The effects of nutritional status on productivity as assessed in certain biomedical studies

The physical work capacity of individuals is measured in biomedical studies via the maximum volume of oxygen consumed during treadmill and other endurance tests (Astrand and Rodahl, 1977). Higher oxygen uptake is an indicator of the ability to perform strenuous tasks. Many studies have been conducted in developing countries and have shown that anthropometric indicators such as height and weight are positively associated with physical work capacity (Spurr, 1983). These are reasonable findings since an under-nourished individual is likely to have low muscle (or fat-free) mass, which in turn limits aerobic capacity. However, there are other nutrients, such as iron, that affect individuals' physical work capacity. For example, while anemic individuals may be able to perform physically demanding tasks, the rate of performance is likely to be slow because low hemoglobin concentration reduces oxygen transportation (see e.g. Gardner *et al.*, 1975). Furthermore, under-nourished individuals may sometimes be forced into undertaking strenuous work, thereby increasing short-run energy expenditures. For example, it is common in developing countries to observe higher labor input levels in certain seasons even if this may entail weight loss. Of course, it is important that body stores of energy and vital nutrients are replenished without individuals succumbing to diseases due to excessive stress. Social scientists interested in issues of labor productivity need to be familiar with some of the evidence from the biomedical sciences in order to interpret results from individual-level studies of nutrition and productivity.

Iron supplementation and the productivity of rubber plantation workers in Indonesia

A well-known randomized controlled trial on the effects of iron supplementation on the productivity of farm workers was conducted by Basta *et al.* (1979) in a rubber plantation in Indonesia. Anemia is common in countries where consumption of animal foods is low. In fact, approximately half the workers in the rubber plantation were anemic (hemoglobin <13.0 g/l for men). While staple foods such as rice and wheat contain ample quantities of non-heme iron, iron absorption rates are low (1–5%) due to the presence of phytates (Bhargava, Bouis *et al.*, 2001). On the basis of blood analyses, 153 of the 302 rubber plantation workers (rubber tappers and weeders) fell into the anemic group, while 149 were placed into the non-anemic group. The workers were randomly assigned to intervention and control groups. The workers in the intervention group received 100-mg iron (ferrous sulfate) tablets for 60 days, while those in the control group received an identical-looking placebo containing tapioca and saccharine. The study was double-blind: neither the workers nor those giving the tablets knew the contents. Moreover, the workers were offered US$0.03 per day to encourage participation, though 41 workers dropped out from the study despite the payments. Another ten workers were dropped from the data analysis because they received medications that might have interfered with the interpretation of the results.

The outcome variables in this prospective study consisted of the Harvard Step Test, which is a simple test of physical work capacity; hemoglobin and hematocrit measurements for iron status; and productivity measures, such as kilograms of wet latex collected by tappers and the area excavated by weeders. Furthermore, intermediate outcomes such as the morbidity and dietary intakes of workers were also monitored, in part to minimize the influence of factors that can obscure the effects of improved nutritional status on labor productivity. For example, the small payments to both groups were found to increase the consumption of vegetables that are good sources of vitamins A and C and, in turn, increase the absorption of non-heme iron and hence hemoglobin concentration (Bhargava, Bouis *et al.*, 2001).

The main conclusions from this randomized controlled trial in Indonesia were as follows. First, there were significant increases in the hemoglobin concentration of anemic workers after the intervention (P < 0.001). By contrast, the hemoglobin concentration of non-anemic workers did not increase significantly. Moreover, the differences in changes in hemoglobin in the two groups were significant around the 8% level. A subset of workers was tested for intestinal parasites such as hookworm, and 45% had "light" or "heavy" parasitical infections. Intestinal parasites are known to lower iron status, so that randomized trials offering both anthelmintic treatment and iron supplementation are likely to have stronger effects on biological markers such as hemoglobin concentration (Bhargava, Jukes *et al.*, 2003). Second, there were significant differences in latex collected by anemic and non-anemic workers (P < 0.01). Moreover, following the supplementation, the output of anemic workers (often classified by their supervisors as "lazy" or "weak") increased by 15%, which was significant at the 5% level. By contrast, there were only small increases in the output of non-anemic workers, and differences in changes in the two groups were statistically significant. Last, there were significant differences between performances of anemic and non-anemic workers on the Harvard Step Test before the intervention (P < 0.001), and improvements in the performance of anemic workers were significantly greater than those for non-anemic workers (P < 0.05).

Interpreting the findings of randomized controlled trials from a broad policy perspective

While the results from this relatively small randomized trial in Indonesia demonstrated the importance of reducing iron deficiencies for increasing labor productivity, it would be helpful to put the findings in broader methodological and policy contexts. The relationships between iron intakes, hemoglobin concentration, oxygen transportation, and the ability to perform strenuous tasks are clearly understood in the biomedical sciences. Thus, randomized trials might have at least two purposes. First, one may wish to quantify the effects of a supplementation administered over a short time span, such as 60 days, to deduce the benefits for iron status and labor productivity. The results can provide useful guidance, especially if iron supplementation is contemplated for various population subgroups as a long-term strategy. From this viewpoint, the benefits of iron supplementation are likely to outweigh the costs, which are typically very low (approximately US$0.10 per day). However, there are logistical difficulties in reaching populations in remote areas and in ensuring the regular intake of iron tablets on a daily or weekly basis (Casanueva *et al.*, 2006). Thus, it is often necessary to target vulnerable groups such as pregnant and lactating women through public health clinics in developing countries. Moreover, improving diet quality via increasing the iron content of staple foods and encouraging small dairy farming is likely to be an effective long-term strategy for improving the iron status of populations (Bhargava, Bouis *et al.*, 2001).

Second, the benefits of iron supplementation in developing countries may be hampered by parasitic infections in individuals. In order to reduce iron deficiencies, it is important to reduce the load of intestinal parasites such as hookworm and schistosomiasis (Bhargava, 2001a). While this can be achieved in the short run via medications delivered at (say) six-month intervals (Bhargava, Jukes *et al.*, 2003), to decrease the transmission of parasites it is essential to improve sanitation. However, sewage treatment is an expensive intervention; even countries such as China, with per capita gross domestic product of over US$2,500, cannot afford treatment for all its citizens. Thus, cost–benefit analyses of interventions for improving iron status are complex and are influenced by the level of economic development. However, it is estimated that over three billion people suffer from anemia in developing and middle-income countries (UNICEF/WHO, 1999) and these deficiencies entail productivity losses. Hence it is important to tackle iron deficiencies despite the meager resources. To this end, small-scale trials administering anthelmintic treatment and iron supplementation can provide insights into the expected gains in productive activities. Alternatively, one might deduce the likely benefits using data from countries that are similar in geographic and economic characteristics, since the underlying biological relationships are well understood. Thus, in order to formulate policies to alleviate iron deficiencies, the costs of increasing iron intakes and reducing parasitical loads can be compared with the expected benefits from randomized trials or by estimating models using data from household surveys.

6.2 Empirical models for the relationships between nutrition and productivity

The importance of research on the biochemistry of food for the productivity of individuals was recognized in the economics literature by Leibenstein (1957), who argued that higher wages for workers would improve their nutritional status and hence productivity. Further theoretical contributions to this "wage efficiency hypothesis" in the economics literature were made by Majumdar (1959), Mirrlees (1975), Stiglitz (1976), Bliss and Stern (1978) and Dasgupta (1993). The survey by Strauss and Thomas (1998) discusses empirical models in the literature for the relationships between health and productivity. It should be noted, however, that a large proportion of the population in developing countries subsists on agricultural activities, and energy deficiencies are prevalent, especially in Africa. Thus, the link between wages and labor productivity is affected by intrahousehold distribution of food, since earners' intakes need not increase proportionately with wages (Majumdar, 1959). Furthermore, while higher wages ultimately affect productivity via improvements in health (Fogel, 1994), as argued in Chapter 3, the lags underlying the development of health indicators are complex. Such issues need to be incorporated in economic models of nutrition and productivity in order to enhance the value of the empirical results.

For illustrative purposes, Fig. 6.1 summarizes the basic hypotheses underlying the empirical models specified in the literature for relationships

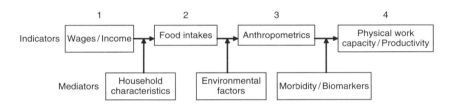

Models for relationships	Studies
2→4	Leibenstein (1957); Mirrlees (1975); Bliss and Stern (1978); Dasgupta (1993)
4→2	FAO/UNU/WHO (1995)
3→4	Bhargava (1997); Thomas and Strauss (1997)[a]
[2+3]→4	Bhargava (1997); Thomas and Strauss (1997)[a]
Long run: 1↔4	Bhargava (2001)

[a] Assumes that wages reflect productivity.

Fig. 6.1 A graphical representation of relationships between income, food, health and productivity in developing countries

between nutrition and productivity. Among the commonly used indicators in the four boxes in the top row, it is postulated that higher wages (or incomes) implies better diet, which, in turn, gradually affects anthropometric indicators such as height and weight. Good nutritional status, reflected in diet quality and anthropometric indicators, affects individuals' physical work capacity (productivity), which is likely to influence wages in the long run. There are also "mediating" variables influencing the relationships in Fig. 6.1. For example, the relationship between wages (or incomes) and food intakes is influenced by household characteristics such as family size and maternal nutritional knowledge (see Chapter 2). Similarly, environmental factors such as water contamination and poor sanitation can hamper the absorption of nutrients, thereby reducing the benefits of nutrient intakes for anthropometric indicators (Chapter 3). Last, sickness spells and nutrient deficiencies can weaken the relationships between anthropometric indicators and physical work capacity. For example, because strenuous tasks require higher oxygen consumption per unit of time, anemic individuals may be unable to perform such tasks even if their height, weight and food intakes appear to be adequate (Bhargava and Reeds, 1995).

The lower half of Fig. 6.1 attributes the main hypotheses linking incomes, food intakes, anthropometric indicators and physical work capacity to contributions in the literature. For example, Leibenstein (1957) emphasized the beneficial effects of higher food intakes for productivity; subsequent work by Mirrlees (1975), Bliss and Stern (1978) and Dasgupta (1993) incorporated other aspects. However, the influence on wages exerted by local labor market conditions and economic activities in the region can complicate the interpretation of the empirical evidence. Thus, for example, Bliss and Stern (1978: 392) reported wide variation in wages within Indian states and, more significantly, within districts and between neighboring villages. Differences in wages between neighboring villages may be partly due to different levels of economic activity and also to the perceived productivity of individuals, reflected in anthropometric indicators (see e.g. Deolalikar, 1988). Moreover, employers concerned with the adequacy of workers' energy intakes for performing strenuous tasks may offer meals during work hours, in part to reduce the length of work stoppages. Such factors are important in practical situations and empirical models inspired by the wage efficiency hypothesis need to be interpreted more broadly.

In contrast with the emphasis in economics literature on the impact of higher food intakes on labor productivity, the nutrition literature summarized in FAO/WHO/UNU (1985) emphasized the need for incorporating individuals' energy expenditures (activities) into the assessment of their energy requirements, i.e. higher food intakes are the *consequence* of higher energy expenditures. Certain issues in interpreting energy intakes/expenditures data were discussed by Bhargava and Reeds (1995). In particular, it was underscored that at low levels of nutrition in developing countries, energy intakes are likely to drive energy expenditures. However, this relationship is reversed in affluent settings, where food consumption is not restricted by poverty and hence energy intakes should be guided by energy expenditures. Overall, it is somewhat simplistic to view energy intakes–expenditures relationships in a short time frame, since even under-nourished individuals have large energy stores that can be readily drawn upon. More importantly, micronutrient intakes, such as those of minerals and vitamins, are critical for maintaining individuals' long-term health and affect labor productivity, especially in developing countries.

Policy implications of the empirical evidence on the links between nutrition and productivity

A survey by Spurr (1983) summarized several biomedical studies linking individuals' anthropometric indicators and iron status to higher physical work capacity and labor productivity. Some of the evidence influenced the specification of econometric models for time allocation patterns in Rwanda (Bhargava, 1997); which is discussed in section 6.3. Moreover, one can view the models for wages estimated by Thomas and Strauss (1997) in this spirit in so far as wages reflect actual productivity, which is likely to be the case provided that heterogeneity levels in economic activity across regions are low. Last, the models estimated by Bhargava (1997) and Thomas and Strauss (1997) also included individuals' intakes of energy and protein as explanatory variables, and were therefore generalizations of models in the biomedical literature. While the long run relationships between health and wages are likely to be bi-directional, as indicated in Fig. 6.1, analyses of data on wages, food intakes, anthropometric measures, morbidity and productivity can afford useful insights on food policy issues (Bhargava, 2001a).

The empirical evidence from the economics literature on nutrition, health and wages was summarized by Strauss and Thomas (1998). For example, Thomas and Strauss (1997) found that wages in Brazil were strongly influenced by individuals' height and weight. Certain models also yielded significant effects of energy and protein intakes on wages.

Because height is a good indicator of childhood nutrition, height can convey useful information to employers regarding the expected productivity of individuals. A positive relationship between anthropometric indicators and wages is likely to be apparent, in so far as there is flexibility in wage offers on the basis of perceived or actual productivity. However, differences in economic activity in different regions can attenuate such relationships. While education can diminish the effects of body size on wages, especially in skilled occupations, it is often the case that educated individuals in developing countries have grown up in better-off households and have received better childhood nutrition, i.e. such individuals are likely to be taller and earning higher wages in skilled occupations. Although one can account for education levels in empirical models, it is important that an unduly high level of heterogeneity in the data not encourage spurious associations between health indicators and wages; longitudinal data can circumvent these problems by controlling for unobserved heterogeneity via the introduction of individual specific random or fixed effects in empirical models.

In view of the evidence in the biomedical sciences on the beneficial effects of good nutritional status on physical work capacity, it would seem important for economists to focus on formulating food policies that enable individuals in developing countries to achieve optimal nutritional and health status given the constraints on resources. This is important, since it may not be rational for employers to offer higher wages to undernourished individuals because of likely delays in improvements in health status and productivity. For example, in countries with energy deficiencies, public food supplementation programs such as those offering "food for work" should be given high priority. Moreover, because iron deficiencies are prevalent in developing countries, it would be useful to increase the iron content of foods that are provided through such programs. Cost–benefit analyses can assess the extent to which energy and nutrient deficiencies can be alleviated by policies such as improving the iron content of staple foods, subsidizing small-scale dairy farms to increase the intake of iron and protein from animal sources, reducing parasitical infections and distributing iron tablets. Development of cost-effective policies may sometimes require randomized controlled trials for quantifying the effects of alternative strategies. However, a comprehensive set of interventions typically cannot be investigated via randomized trial because of the high cost. In practice, researchers have to rely on analyses of data from household surveys, which provide useful insights in most circumstances. Such issues were also underscored in Chapter 4 in the context of randomized controlled trials for treating children against parasitical infections.

In summary, the economics literature has developed many theoretical constructs for analyzing the effects of nutritional status on labor productivity and wages, though these models do not fully incorporate the evidence from biomedical studies. However, the recent emphasis on randomized controlled trials in economics is likely to lead to some convergence of views, in part because the pathways through which nutritional status affects physical work capacity are well understood. Thus, economists involved in formulating food policies are increasingly likely to focus on issues such as the cost-effectiveness of policies to achieve better nutritional and health status in a population. Moreover, as emphasized in Chapter 5, issues of high fertility are intertwined with poor health status, especially of women, and children born at high birth orders. Thus, providing basic health care services to poor households is also likely to improve children's nutritional status, which is important for their future productivity in both skilled and unskilled occupations.

6.3 An analysis of time allocation patterns of adults in Rwanda

Analyses of time allocation patterns of adults in developing countries can afford insights into the effects of nutritional and health status on various activities. As noted in section 6.2, wages earned by individuals are affected by macroeconomic factors, which are seldom incorporated in household surveys. Moreover, one is likely to observe behavioral changes in individuals consuming inadequate quantities of dietary energy, such as spending a greater proportion of time on sedentary activities (e.g. Beaton, 1984; Waterlow, 1986). Thus, it is important to combine the approaches in the biological and social sciences for analyzing time allocation data. Moreover, many adults in developing countries work on their own land and perform housework without receiving monetary compensation. Hence, elaborate data on time allocation patterns, such as those from Rwanda, can provide valuable information on the proximate determinants of productive activities.

Modeling the relationship between nutritional status and time allocation patterns in countries such as Rwanda presents several difficulties. First, while recording daily activities at a disaggregated level is insightful, there is often large internal variation in the data. Second, energy expenditures, which are estimated as multiples of the basal metabolic rate (BMR, the minimal energy necessary for sustaining life), are available for approximately 25 broad categories of activity (FAO/UNU/WHO, 1985). Third, individual food intakes are difficult to quantify in many sub-Saharan African countries because household members share food from the same plate. Fourth, alternative formulations for sharing work between men and women are available in the social sciences (Becker, 1965, 1991; Sen, 1983; Berio, 1984). However, some of the assumptions underlying these formulations remain untested and simpler approaches to the labor supply of men and women (Mincer, 1962; Ashenfelter and Heckman, 1974) can provide a flexible framework. Last, due to differences in data collection procedures, the effects of variables such as food prices on time allocation patterns can be assessed only indirectly in a longitudinal framework. These issues are addressed in the next subsection using a unique data set from Rwanda (Bhargava, 1997).

The longitudinal data set from Rwanda

The longitudinal study in Rwanda was conducted during 1982/3 (République Rwandaise, 1986). Food intakes in 270 households (three from each of the 90 "sectors") were recorded for seven consecutive days. The surveys were repeated four times at four-month intervals. Heights of adults were recorded once and weights were measured in each round. The time spent by household members on around 900 activities for 14 consecutive days was recorded in each survey round. Since food intakes were observed for seven days, activity data on the same days were used in the analyses (Bhargava, 1997). Further, to reduce internal variation in the data, activities were mapped into 25 broad categories defined by FAO/UNU/WHO (1985). Activities that were difficult to match were put into five groups constructed on the basis of energy expenditures (expressed as multiples of BMR). The data for seven days were then averaged to produce a figure for an "average" day of the week, i.e. average activity levels in the four survey rounds were analyzed.

Information on education, landholding and other variables was available in the data set. Single observations for the survey period were available on the value of households' consumption and production. The prices of seven food groups and information on wages earned by a subset of household members were recorded at one point in time during the survey. The "lead" young adult man and woman were selected from each household; these adults may not have been a couple but were likely to be the ones performing demanding tasks. Retaining individuals with four time observations, complete data were obtained on approximately 110 adult pairs. The sample means are reported in Table 6.1. Average energy expenditures for men and women were around

Table 6.1 Sample means of the variables in the four rounds of data from Rwanda

Variable	Men	Women
Proportion of time spent sleeping, resting and sitting quietly	0.547	0.509
	(0.107)	(0.086)
Proportion of time spent on heavy activities[a]	0.187	0.180
	(0.088)	(0.073)
Proportion of time spent on housework	0.014	0.131
	(0.050)	(0.069)
Proportion of time spent on agriculture	0.100	0.140
	(0.088)	(0.079)
Average energy expenditure[b]	1.992	2.184
	(0.437)	(0.298)
Household size	5.991	
	(2.248)	
Age	34.223	32.017
	(19.315)	(16.157)
Total value of household consumption[c]		51,881
		(25,319)
Total value of household production[c]		43,271
		(31,248)
Household energy intake (kcal/day)		11,161
		(5,340)
Household protein intake (g/day)		361.6
		(201.7)
Weight (kg)	48.817	48.269
	(12.787)	(11.636)
Height (cm)	157.259	152.975
	(15.625)	(11.637)
Body mass index (BMI)	19.350	20.333
	(3.033)	(3.592)

Note: Sample means are calculated using 4 time observations on 116 men and 119 women; numbers in parentheses are standard deviations.
[a] Heavy activities require at least three times as much energy as sleeping.
[b] Expressed in terms of Basal metabolic rate.
[c] Annual figures in Rwandese francs.
Source: Bhargava (1997).

twice the BMR and hence these adults were physically very active (Schofield, 1986) despite poor nutritional status, reflected in the BMI means.

Modeling the effects of economic variables on time allocation patterns

The treatment of leisure (L) as an argument in the utility function in economic analyses yields the demand for leisure as a function of goods prices (p), wage rate (w) and non-labor income (A):

$$L = h(p,w,A) \qquad (6.1)$$

However, it is desirable to adopt a more flexible approach in the analysis of longitudinal time allocation data from Rwanda. First, a large proportion of food consumed by households was their own produce or received in the form of gifts. Also, energy deficiencies were apparent in Table 6.1 from the sample means of BMI, which were 19.4 and 20.3 for men and women, respectively (James *et al.*, 1988). In such circumstances, it was likely that energy expenditures were driven by energy intakes and not the converse (Bhargava and Reeds, 1985). Thus, the food available to many households was inadequate and could be viewed as pre-allocated (Pollak, 1969). Moreover, indicators of nutritional status such as BMI were potentially important explanatory variables in time allocation models, since they reflect the effects of current and past nutritional deficiencies. Second, due to self-production of food and gifts received, minor fluctuations in food prices may not have immediate effects on leisure activities, though prices will gradually influence "habitual" levels of leisure activities. Third, since a single observation on wages was available in the data for roughly a third of the households, it might seem difficult to examine the effects of wages on time allocation patterns. However, one can control for the earnings of household members by including total consumption during the sample period as a regressor in longitudinal models. The link between nutritional status and time allocation was investigated using the data set from Rwanda by splitting the estimation strategy into two stages.

In the first stage, households' average energy intakes in the four survey rounds were explained in a cross-sectional framework by variables such as household size, regional dummy variables, food prices and measures of household incomes. The estimated relationship provided a large-scale view of the effects of economic variables on food availability. The second stage modeled the determinants of time allocated to leisure and productive activities in a longitudinal framework. Since the choice between work and leisure is likely to be affected by health, biological and behavioral factors played an important role in model specification. In particular, the mechanism by which energy deficiencies restrict activities, and factors underlying the sharing of work between men and women, were addressed. For example, women with young children may substitute housework for agricultural activities. Moreover, inadequate energy intakes may force reductions in the energy expenditures of adults. This might be achieved by decreasing the effort on strenuous tasks such as agricultural activities. If energy deficiencies persist over time, then individuals will lose weight, as was observed in these longitudinal data (Table 6.1).

Results for average energy intakes in Rwandese households

Table 6.2 presents the results for households' average energy intakes in the four survey rounds. Since Rwanda is divided into six geographical zones, five dummy variables were included in the models; dummy variables for zones 3, 5 and 6 were insignificant. The results in the first two columns assumed the regressors to be exogenous; the third column treated the total value of consumption as an endogenous variable and reports instrumental-variables estimates. All variables were transformed into natural logarithms to reduce heteroscedasticity.

Household size was positively and significantly associated with energy intakes (P < 0.05). While some non-linearity was implicit in the logarithmic specification, the square of household size was insignificant. The total value of consumption was an approximate measure for household incomes. The point estimate of the income elasticity of energy intakes in the first column

Table 6.2 Cross-sectional results for average household energy intakes, Rwanda

Variables	Specification 1[a]	Specification 2[a]	Specification 3[b]
Constant	6.476*	6.204*	5.376*
	(0.509)	(0.661)	(1.909)
Zone 2[c]	−0.119*	−0.107	−0.110*
	(0.041)	(0.062)	(0.045)
Zone 4[c]	−0.186*	−0.197*	−0.179*
	(0.041)	(0.059)	(0.044)
Household size	0.234*	0.253*	0.154
	(0.041)	(0.061)	(0.139)
Total consumption	0.558*	0.545*	0.667*
	(0.040)	(0.053)	(0.182)
Total production	0.002	−0.006	
	(0.022)	(0.032)	
Average wage rate		−0.047	
		(0.032)	
Price of beans	−0.146	−0.093	−0.120
	(0.091)	(0.130)	(0.102)
Price of sweet potatoes	−0.095*	−0.020	−0.101*
	(0.046)	(0.062)	(0.048)
Price of traditional beers	0.477*	0.568*	0.476*
	(0.077)	(0.114)	(0.077)
Sample size	251	119	251
Adjusted R^2	0.757	0.721	

Notes: All variables are in logarithms; standard errors are in parentheses.
[a] Estimated by least squares.
[b] This specification treats consumption as endogenous using land size as an additional instrument.
[c] Indicator variables.
*P <0.05.
Source: Bhargava (1997).

was 0.56. The total value of production, however, was insignificant. Because energy intakes were likely to influence productive activities, consumption could be correlated with the error terms. This problem was tackled in the third column, where size of land owned by the household was used as an additional instrumental variable in the estimation. Landholding is fixed in Rwanda by the government and there was much variation in land quality. The estimate of income elasticity was higher in the third column (0.67), though its standard error was larger as well.

Higher prices of sweet potatoes significantly decreased households' long-run energy intakes. Due to poor land quality in Rwanda, some quantities of staple foods are purchased at market prices. The price of traditional beers had the opposite effect on energy intakes. Thus expenditures on beer (consumed mainly by men) appeared to divert scarce resources from staple foods. Last, the results in the second column included a measure for wages earned by household members; the wage variable was insignificant. The reduction in sample size decreased the precision of estimates and prices of beans and sweet potatoes were insignificant as well. The treatment of wages and consumption as endogenous variables led to insignificance for all the regressors. There were difficulties in finding suitable instruments for predicting wages. Moreover, as mentioned above, measures of macroeconomic activity in 90 sectors of Rwanda were not included in the data set.

Results for resting and sleeping patterns in Rwanda

The results for proportion of time spent by adults resting, sleeping and sitting quietly are in Table 6.3; logistic transformation of the dependent variable ensured a smooth relationship with explanatory variables. Height and weight were initially introduced as separate regressors. The use of likelihood ratio statistics, discussed in Chapter 3, generally led to acceptance of the hypothesis that these two variables can be combined as the BMI. Moreover, in models where coefficients of height and weight were unrestricted, collinearity among the regressors was evident.

Focusing first on the results for men, time spent resting increased with age, though the coefficient of age squared was insignificant. Also, the resting time of the lead adult was positively associated with household size. Second, estimated coefficients of the value of household consumption and individuals' BMI were negative and significant. The coefficient of BMI was large and this variable reflected the cumulative effects of energy deficiencies. It appeared that adult males in poor households spent additional time resting

Table 6.3 Quasi-maximum-likelihood estimates for the proportion of time spent sleeping, resting and sitting quietly, Rwanda

Variable	Men	Women
Constant	−0.915	1.387
	(0.876)	(1.356)
Age	0.129*	−0.006
	(0.044)	(0.047)
Household size	0.149*	0.095
	(0.075)	(0.078)
Total consumption	−0.308*	0.038
	(0.093)	(0.078)
Protein intake	0.042	0.070
	(0.075)	(0.069)
Average	−0.316*	−0.165
energy expenditure[a]	(0.163)	(0.094)
Body mass index (BMI)	−0.623*	0.025
	(0.161)	(0.139)
Energy intake	−0.031	−0.174
	(0.102)	(0.105)
Time dummy 3	0.211*	0.153*
	(0.060)	(0.044)
Time dummy 4	0.121*	0.117*
	(0.067)	(0.047)
Lagged dependent variable	0.322*	0.021
	(0.095)	(0.083)
Between–within variance	0.007	0.364*
	(0.063)	(0.163)
Within variance	0.192	0.090
Chi-square (12)[b]	15.8	18.8

Note: The dependent variable is the logistic transformation; standard errors in parentheses.
[a] Women's average energy expenditure is included in the equation for men and vice versa.
[b] The test for the exogeneity of men's (or women's) average energy expenditure, BMI and energy intakes.
*$P < 0.05$.
Source: Bhargava (1997).

and sleeping to avoid weight loss, which was also evident between the first and fourth survey rounds (Table 6.1). Third, the time spent resting by men was negatively associated with the average energy expenditure of the lead woman in the household. Since household consumption was taken into account, a possible explanation was that men decreased their resting time to offset high demand for women's activities. This demonstrated the joint nature of activity patterns; a decline in men's leisure would lessen the work-load of women on subsistence tasks. While the type of work undertaken is explored below, the marginal product of labor was likely to be low due to poor nutritional status. However, no further measures of adult productivity were available in the data set.

Fourth, energy intakes were statistically insignificant in the model for time spent resting, sleeping and sitting quietly. A likelihood ratio test accepted the joint exogeneity of women's average energy expenditures and of men's BMI and energy intakes. However, if the latter two variables were treated as exogenous, the exogeneity of women's (men's) expenditures was rejected in some of the models. Last, the coefficients of indicator variables for the third and fourth survey rounds were positive and significant, indicating low work availability. Also, time spent resting was significantly influenced by its lagged value; the long-run effect of a change in an independent variable was about 1.33 times the short-run impact. The between-subject variance was insignificant in the model for time spent resting, sleeping and sitting quietly.

In contrast with the results for men, only a few variables were significant in explaining women's resting patterns. Average energy expenditure of men and women's energy intakes were estimated with negative coefficients that were significant at the 10% level. The indicator variables for survey rounds and the between-subject variance ratio were significant. Note that the estimated within-subject variance was twice as large for men. This might be due to the fact that agricultural tasks performed by men were seasonal in nature. In contrast, women were expected to perform subsistence activities throughout the year.

Results for agricultural and household activities in Rwanda

The results for time allocated to agricultural and household activities (with a continuity correction for extreme values; see Cox, 1970) are presented in Table 6.4. The models underscored the joint nature of time allocation decisions; men's agricultural activities depended on their effort on housework and on time spent by women on agriculture. Women's housework was influenced by their agricultural activities. In contrast, women's agricultural activities were influenced by their housework and by the time spent by men on agriculture. First, in the results for men's agriculture, age, total value of consumption, BMI, energy intakes, women's agriculture, time dummies and between-subject variance were all significant predictors. The coefficient of age was positive and inclusion of squared age indicated a non-linear relationship, though the squared term was not significant at the 5% level. Also, housework performed by men was insignificant.

In the model for women's housework, the coefficient of household size was significant; a greater number of household members was associated with reduction in the lead woman's effort on housework. This phenomenon was explored further by controlling for numbers of children in age

Table 6.4 Quasi-maximum-likelihood estimates for the proportion of time spent on agricultural and household activities, Rwanda

Variable	Men	Women	
	Agriculture	Housework	Agriculture
Constant	0.858	0.917	1.635
	(1.384)	(2.496)	(1.763)
Age	0.331*	0.139	0.518*
	(0.107)	(0.094)	(0.097)
Household size	−0.148	−0.360*	−0.116
	(0.164)	(0.155)	(0.154)
Total consumption	0.149*	−0.076	0.234
	(0.067)	(0.165)	(0.163)
Protein intake	−0.203	−0.070	−0.084
	(0.120)	(0.116)	(0.126)
Men's or women's agriculture[a]	0.239*	−0.113*	0.153*
	(0.062)	(0.049)	(0.041)
Housework[b]	0.032		−0.152*
	(0.090)		(0.060)
BMI	0.937*	0.773*	0.773*
	(0.327)	(0.247)	(0.194)
Energy intake	0.226*	0.253	−0.134
	(0.115)	(0.166)	(0.181)
Time dummy 3	−0.117	−0.122	−0.150
	(0.091)	(0.065)	(0.072)
Time dummy 4	−0.176*	−0.089	−0.023
	(0.090)	(0.067)	(0.075)
Lagged dependent variable	0.068	0.009	−0.027
	(0.081)	(0.067)	(0.074)
Between–within variance	0.633*	0.802*	0.624*
	(0.233)	(0.278)	(0.225)
Within variance	0.427	0.208	0.256
Chi-square[c]	18.7 (16)	7.7 (12)	32.8 (16)

[a] Men's agriculture appears in the equation for women's agriculture and vice versa.
[b] Women's agriculture is included in the women's housework equation.
[c] Degrees of freedom of the chi-square test for the exogeneity of agriculture, housework, BMI and energy intakes are in parentheses.
* $P < 0.05$
Source: Bhargava (1997).

groups 1–5 and 5–10 years. These estimates were less precise, as the sample size was reduced to 85 women because of missing data. However, the results confirmed that children in households enabled women to increase their resting time (Schultz, 1973). The coefficients of women's agriculture and BMI were significant in explaining housework, which contrasted with the findings in Table 6.3. Household and agricultural activities seemed to cover the salient aspects of women's role in subsistence agriculture. Moreover, there was substitution between agricultural and household activities for women, though this was not the case for men.

Finally, the results for women's agriculture were broadly consistent with the estimates obtained for housework. The coefficient of men's agriculture was significant and positive, whereas that of housework was negative. Substitution between agricultural and household activities was of similar order of magnitude in the two models. BMI was positively associated with agriculture and household size was insignificant. The coefficients of dummy variables for the third survey round were negative and significant in the two columns. The coefficients of lagged dependent variables were insignificant, though the between-subject variances were significant. More generally, it was evident from these results that men with higher BMI spent a greater proportion of time on agricultural activities, whereas women with higher BMI spent a greater proportion of time on housework as well as agricultural activities. Thus, these estimates supported the basic hypotheses linking better nutritional status to higher labor productivity. Note that these results were unlikely to be contaminated by "reverse causality", since higher levels of household and agricultural activities are unlikely to raise adults' BMI.

6.4 Conclusion

This chapter described some important aspects of the relationships between nutrition, health and labor productivity in developing countries. With improvements in health status due to economic development, one is likely to see an increase in productivity levels. The evidence from biomedical studies clearly indicates beneficial effects of nutritional status on physical work capacity and labor productivity. Moreover, randomized controlled trials of iron supplementation have shown increased productivity, especially among anemic individuals. Thus, it would not seem particularly important to replicate these studies in other countries. Instead, it is critical to devise food policies that improve individuals' nutritional and health status and raise productivity levels via cost-effective food supplementation programs.

There has been some emphasis in the economics literature on the effects of nutrition and wages on productivity. Because higher intakes of desirable nutrients gradually improve individuals' health status, employers are likely to offer higher wages to workers that appear to be well nourished and may not offer similar terms to under-nourished individuals. Thus, it is important for national and international agencies to devise food supplementation programs for under-nourished individuals in developing countries. The data from Rwanda indicate the importance of nutritional status for productive activities. Unlike wages, which are influenced by many macroeconomic factors, time allocation patterns provide direct information on time spent on productive activities. It was seen in the Rwandese data that adults with energy expenditures around twice the BMR lost weight, presumably due to the high workload. Thus, food supplementation programs are likely to enhance health status and labor productivity. Deficiencies of various micronutrients can be identified from nutritional surveys and economists can play an important role in devising food policies for raising labor productivity in developing countries. For example, increasing the intakes of protein and iron, especially from animal sources, are likely to enhance the performance of agricultural and heavy activities. Improvements in health status and life expectancy are important predictors of GDP growth rates in low-income countries (Bhargava, Jamison *et al.*, 2001) and hence investments in health are likely to have high payoffs in the long run.

7

Behavior, diet and obesity in developed countries

The final chapter of this book is concerned with issues of individual behavior, diet, and obesity, especially in developed countries such as the US. The discussion provides an interesting contrast to earlier chapters, which underscored the importance of good nutrition for child health and for the labor productivity of adults in developing countries. The obesity epidemic also affects affluent sections of societies in middle- and low-income countries, so that these issues are of global relevance. With increased affluence among populations, food expenditures are a small proportion of households' total budgets. Moreover, food prices are low partly because of agricultural subsidies in developed countries, and prices in fast-food restaurants have declined due to low wages in the service sector. In addition, the opportunity to sample various cuisines can promote over-eating. Thus, individuals with poor dietary knowledge and self-control are likely to over-consume food, and these problems are compounded by the fact that higher body weight, in turn, increases individuals' energy requirements. As discussed in preceding chapters, individuals' basal metabolic rate (BMR) is positively predicted by body weight, and energy required for various physical activities increases with BMR. Maintaining a healthy body weight is a challenge among affluent populations; it is perhaps not surprising that two-thirds of the US population is overweight, i.e. with body mass index (BMI) greater than 25 (World Health Organization, 2006). Costs of treatment of chronic conditions associated with obesity, such as high blood pressure, diabetes, cardiovascular disease and cancers, are high. Moreover, in developing countries, chronic diseases result in a perverse reallocation of medical resources: "traditional" diseases of poverty are neglected because treatments of chronic diseases are lucrative

(Bhargava, 2001b). A preventive approach to chronic diseases via healthful eating and physical activity is a sound long-term strategy for improving a population's health.

Economists and social scientists interested in obesity issues analyze data from surveys that gather information on individuals' weight, socio-economic and demographic variables, food consumption and shopping patterns, neighborhood characteristics such as access to parks and types of restaurants in the area, etc. In contrast, nutritionists and epidemiologists are concerned with biological pathways through which food intakes might lead to weight gain. Psychologists working in obesity research are interested in factors, such as individuals' motivation to change lifestyle, beliefs about health and degree of self-efficacy (Bandura, 1977; Rosenstock *et al.*, 1986) that can facilitate weight loss. While alternative approaches to obesity in different fields provide useful insights, it is important to recognize their relative strengths and weaknesses. For example, National Health and Nutrition Examination Surveys (NHANES; National Center for Health Statistics, 2007) collect useful dietary and anthropometric information on large samples ($\sim$5,000) of individuals in the US. Due to the cross-sectional nature of these surveys, however, evidence on the effects of dietary intakes on obesity is often not compelling. For example, one might find positive associations between individuals' fat intakes and BMI that could be due to higher fat intakes leading to higher body weight. However, the converse of this phenomenon is also possible, i.e. heavier individuals consume energy-dense foods such those high in fats because their energy requirements are higher. It is evident that analyzing the proximate determinants of obesity entails understanding the literature in different fields and examining the relative merits of various approaches.

Further, analyzing the underlying determinants of obesity presents methodological challenges because different disciplines pursue different types of analyses. For example, economists are concerned with the effects on weight gain of low prices of foods such as those high in sugars and fat, and may want to propose schemes such as taxing "junk" foods. Moreover, the availability of energy-dense foods in fast-food restaurants, especially in poor neighborhoods, can encourage overeating. However, evidence based on food availability and obesity prevalence rates will generally be insufficient for devising tax policies, since food consumption is voluntary and individuals derive pleasure from eating. Furthermore, this type of evidence is often unacceptable for nutritionists and epidemiologists because the effects of food intakes on measures of obesity such as BMI or waist-to-hip circumference ratio cannot be demonstrated without elaborate studies. Because weight gain is a gradual process resulting from energy intakes higher than necessary, nutritionists and epidemiologists conduct longitudinal studies monitoring individuals over time. While the biomedical approach is driven by specific scientific hypotheses, even moderately small studies with (say) 100 individuals are expensive. Because it is desirable to have large variation in dietary intakes to assess the effects of food intakes on obesity measures, randomized controlled trials are often conducted with appropriately defined control and intervention groups.

The gradual increases in body weight start from intra-uterine development and are influenced throughout life by a variety of environmental factors. While environmental factors interact with genetic polymorphisms, in view of the trends in obesity prevalence, environmental factors such as maternal nutritional status, eating patterns at home and in restaurants during childhood and in adult life, and physical activity are likely to be key determinants of body weight. In most applications, however, one cannot obtain information on all of the relevant factors affecting body weight. Instead, data on individuals collected during certain time intervals are analyzed to explain differences in anthropometric measures by dietary, economic and psychological factors. Even so, certain variables are more relevant for explaining obesity measures, depending on the time frame of the studies. For example, the relationships between food intakes and obesity measures postulated in the nutritional-epidemiologic literature are likely to be influenced by food prices and incomes if the studies cover a long time frame. However, in nutritional studies the effects of food prices may not be evident in short time intervals. Thus, analyses in different disciplines implicitly employ different time frames and recognizing these aspects is helpful for understanding the determinants of obesity. Section 7.1 begins with nutritional issues and discusses the importance of hypotheses, such as that excessive fat intakes lead to weight gain, and summarizes some relevant studies. Section 7.2 describes salient aspects of the psychological literature and their relevance for designing interventions that can facilitate behavioral changes for promoting weight loss. Section 7.3 briefly discusses the economics literature. Last, Section 7.4 highlights ways in which biomedical and social science research can be integrated to develop strategies for stemming the obesity pandemic.

7.1 Nutritional and epidemiological approaches to the prevalence of obesity

A fundamental tenet of the literature on human physiology is that energy intakes higher than expenditures will lead to weight gain (e.g. Goldberg *et al.*, 1991). The fact that many individuals in developed countries are overweight suggests that it is relatively easy to get trapped into positive energy disequilibrium. Moreover, certain types of diets and behaviors can make individuals susceptible to overeating. While energy absorption rates from various types of foods are largely unknown, diet composition can potentially play an important role in promoting weight gain. For example, laboratory experiments have shown that the human body does not maintain its fat balance as accurately as its carbohydrate and protein balances (Flatt, 1987). Thus, consumption of foods high in fats can more easily lead to weight gain than foods high in carbohydrates and protein. However, this phenomenon has not been rigorously demonstrated at the population level and such issues will be discussed in section 7.1.2. Because of problems in assessing individuals' "habitual" food intakes, commonly used dietary assessment methods in the fields of nutrition and epidemiology are described in section 7.1.1.

7.1.1 *A brief review of dietary assessment methods*

The assessment of individuals' "habitual" intakes of energy and nutrients is complex, in part because of the nutrient composition of foods and also because dietary assessment methods can themselves induce changes in consumption behavior (Block, 1982; Willett, 1998b). The methods commonly used in developed countries are the food frequency questionnaire (FFQ), 24-hour recalls, and food records kept by individuals over three, four or seven days. The FFQ method is widely used: enumerators or individuals fill in responses to items such as how frequently milk, chicken, bread, etc. were consumed in the last 90 (or fewer) days. The numbers of food groups covered are typically not very large (e.g. 200) and the responses are recorded on machine-readable forms. Thus, the time spent filling in an FFQ is short, and food intakes are converted into energy and nutrient intakes quickly and cheaply. However, the availability of thousands of foods, especially to consumers in developed countries, means that assessments of intakes using an FFQ may be less accurate than desired from a research standpoint. Nevertheless, dietary patterns over an extended period that are assessed via FFQs provide important insights into individuals' "habitual" intakes.

The 24-hour recall method was previously discussed in Chapter 2 as a way of measuring food intakes in developing countries. It is also a useful tool in developed countries, since food consumption recorded for the previous day is not affected by possible changes in behavior during the observation period. However, there is often a fair amount of intra-individual (within-subject) variation in intakes, since eating patterns may depend on the day selected for 24-hour recall surveys. i.e. a "typical" day may not reflect "habitual" intakes. Some of these difficulties can be ameliorated via "repeated" 24-recall surveys, i.e. by conducting unannounced 24-hour recall surveys two or three times (Buzzard *et al.*, 1996). Furthermore, individuals tend to forget some of the foods consumed in a previous 24-hour period and hence energy and nutrient intakes assessed from a single 24-hour recall may be lower than those from multiple 24-hour recall surveys or from other methods. The numbers of food groups are generally limited in 24-hour recall surveys and it is difficult to assess food portion sizes, especially during telephone interviews, sometimes conducted in developed countries. By contrast, enumerators visit households in developing countries and carry with them utensils to help assess food portion sizes.

Another useful method of dietary assessment is the three-, four- or seven-day food records, where individuals keep a record of everything consumed during the observation period. For example, the Pyramid Serving Database (National Cancer Institute, 2006), consists of 4,542 foods and hence accurate information can be compiled on various types of meats, bread, and fruits and vegetables consumed. Moreover, eating patterns can differ between weekdays and weekends (Bhargava, Forthofer *et al.*, 1994); seven-day food records cover food consumption for the entire week and are therefore likely to provide insights into "habitual" intakes. However, it is expensive to convert food intakes into energy and nutrient intakes because the records are often not machine-readable. More importantly, individuals filling in food records can become conscious of their intakes and change consumption behavior on the days of record-keeping. This may especially apply to "undesirable" nutrients such as fats, whose consumption may be lowered, while the intakes of "desirable" nutrients such as vitamin C can increase (Bingham, 1994). Thus, in spite of the high quality of dietary information in food records, behavioral changes can undermine assessment of "habitual" energy and nutrient intakes. It is important to recognize such features of dietary assessment methods in analyses of nutritional data; specific issues are discussed in section 7.1.3. In general, because of the paucity of studies recording dietary information using alternative methods for a large number of individuals, it is difficult to compare the relative merits of dietary assessment methods. Nevertheless, information on dietary intakes is useful and can be employed to construct averages for population groups and also for individuals, especially when multiple observations are available on food intakes.

7.1.2 *Anthropometric measurements for obesity*

As noted in Chapter 3, increases in the height and weight of children in developing countries reflect different dimensions of growth, since nutritional deficiencies and morbidity are likely to affect these indicators differently. For adults, height is fixed, though it may start a very gradual decline after 40 years of age. The BMI is a commonly used measure, with the condition "overweight" defined as, $BMI > 25$; for an obese person, $BMI > 30$. However, the use of BMI can be complicated in some situations, such as athletes or individuals in physically demanding occupations with large muscle mass, who are likely to have a high BMI but are in fact not overweight. While BMI is a useful outcome variable reflecting obesity, a more general formulation would explain body weight by several explanatory variables, including individuals' heights. This approach was illustrated in Chapter 3, where a likelihood ratio statistic was used to test whether height and weight should be combined as the BMI. A similar statistic can be developed for testing whether the dependent variable should be individuals' weight or BMI. For example, we can write a model for the body weight of N individuals in the sample as:

$$\ln(\text{Weight})_i = a_0 + a_1 X_{1i} + a_2 \ln(\text{Height})_i + u_i \quad (i = 1, 2, \ldots, N) \quad (7.1)$$

where X_{1i} are explanatory variables in the model in addition to individuals' heights. If the null hypothesis $a_2 = 2$ is accepted using a likelihood ratio or a t-test, then it will be appropriate to model individuals' BMI for investigating the proximate determinants of obesity. However, when sample sizes are small, including the additional variable of height, as in equation (7.1), can create difficulties for estimation methods. Moreover, the model for BMI may yield very similar parameter estimates that are more precise. Also, in certain applications, one might test the null hypothesis that $a_2 = 1$ and acceptance of this hypothesis will imply that it is appropriate to model individuals' weight-by-height.

Another indicator based on anthropometric measurements is the ratio of waist-to-hip circumferences, which is often a significant predictor of outcomes in affluent populations such as coronary heart disease (Larsson *et al.*, 1984; Rexrode *et al.*, 1999). As in the case of combining height and weight, one can test whether waist and hip circumferences should be combined into a single explanatory variable using a likelihood ratio statistic. However, from a biological standpoint, there may not be significant differences between types of fat stored around the waist and hips. Thus, anthropometric assessment and the epidemiologic literature generally combines these two measurements as waist-to-hip ratio, and this variable is an important predictor of biological markers such as the cholesterol and insulin concentrations of women (Bhargava, 2006b). Further research on height, weight, and waist and hip circumferences, using data on large numbers of individuals, can provide insights for combining anthropometric measures for predicting disease outcomes.

7.1.3 Food intakes and anthropometric measures for obesity

While energy intakes higher than necessary will lead to weight gain, the composition of the diet is critical for maintaining the balance between energy intakes and energy expenditures. As noted in Chapter 2, a gram of dietary fat contains 9 kcal of energy, whereas a gram of sugar contains 4 kcal. Thus, consuming foods high in fat can easily increase overall energy intake. Moreover, sugary foods, including soft drinks, are a rich source of energy and may not give individuals the feeling of "satiation", thereby leading to over-consumption. In addition, the safety of food products in developed countries is monitored and foods are less likely to be contaminated, so that energy absorption rates are higher than in developing countries. In such circumstances, it is difficult to maintain an energy balance unless individuals judiciously avoid foods high in energy, and maintain a dietary plan to offset the consumption of energy-dense foods with subsequent intakes of low-energy, high-volume foods such as vegetables.

The role of dietary fat in promoting obesity has received some attention in the nutrition literature, in part because clinical studies have indicated that the human body does not precisely regulate the fat balance, especially in less active individuals (Flatt 1987; Schutz et al., 1989; Shepard et al., 2001). The body is capable of oxidizing carbohydrates and proteins quite efficiently but this may not be true for dietary fat. For example, as lifestyles become sedentary and individuals consume higher quantities of fat, especially in fast-food restaurants, one might see an increase in obesity. However, similar weight gain can result if the overall energy intake is consistently higher than expenditures, and high-fat diets are only one of the many ways through which energy intake can increase. Of course, it is relatively more difficult to achieve high energy intake on a diet consisting primarily of vegetables and whole-grain products, though with the availability of foods high in oils and sugars, weight gain is apparent even among vegetarian populations. Overall, the energy density of the diet is likely to be important for avoiding weight gain (Rolls *et al.*, 1999); consuming diets low in energy density which provide an apparent feeling of "satiety" in terms of volume of food can facilitate the maintenance of energy balance.

A major controversy in the nutrition literature is whether dietary fat is responsible for promoting obesity (Bray and Popkin, 1998; Willett, 1998a). This is not an easy issue to address in observational and experimental studies among free-living population groups, for several reasons. First, as noted in section 7.1.1, measurement of "habitual" intakes is complicated, due to such factors as incomplete recording of foods consumed and potential behavioral changes when keeping food records. Such factors complicate the modeling of relationships between anthropometric factors and diet. Second, while experimental studies have been conducted with intervention groups receiving dietary advice (Henderson *et al.*, 1990; Bowen *et al.*, 1999), these are expensive and often cover short periods such as one to two years. While the ongoing Women's Health Initiative (Prentice *et al.*, 2006) covers an eight-year period, food intakes are recorded at six-monthly intervals using the FFQ method. Moreover, factors such as subjects' understanding of their dietary needs and their perceptions of health can affect anthropometric measures in *both* control and intervention groups. Thus, for example, well-educated and motivated women in the control group can alter their behavior and consume healthier diets, thereby diminishing differences between control and intervention groups. In such circumstances, it is important to investigate the proximate determinants of weights in the control and intervention groups, rather than focusing on differences between the groups, as is often the case in the biomedical sciences.

In spite of the difficulties in modeling the effects of dietary intakes on measures of obesity, nutritional and epidemiological studies have compiled numerous data sets to elucidate these relationships. From the standpoint of the effects of dietary fat on obesity, Sheppard *et al.* (1991) is one of the few studies reporting that a lower percentage of energy derived from total fat led to greater reductions in body weight, in this case among 171 subjects in the intervention arm of the Women's Health Trial. However, the analysis did not control for behavioral and socio-economic factors, and estimation techniques did not fully exploit the longitudinal nature of the data; annual changes in body weight were analyzed while observations were available at six-month intervals. More importantly, the different sources of dietary energy, such as that from carbohydrates and protein, were not accounted for in the model. Also, researchers have argued that saturated, monounsaturated and polyunsaturated fats can have different effects on measures of adiposity (Doucet *et al.*, 1999). More comprehensive investigations, using data on larger number of subjects and taking into account some of the above-mentioned problems, can provide useful insights into the effects of dietary intakes on anthropometric measures. Section 7.1.4 describes a study for post-menopausal women in the US which tackles some of the methodological problems (Bhargava and Guthrie, 2002).

7.1.4 *A case study analyzing the effects of women's food intakes on anthropometric measures*

The Women's Health Trial: Feasibility Study in Minority Populations (WHTFSMP) was a multi-center randomized trial in 1991–5 involving 2,208 women in Atlanta, Birmingham, Alabama, and Miami, with a goal of reducing energy intake from fat to 20% in the intervention group (Bowen *et al.*, 1996). The participants (28% black, 16% Hispanic and 54% white) were post-menopausal women in the age group 50–79 years; 40% and 60% of the women were randomly assigned to the control and intervention groups, respectively. Complete longitudinal data at baseline, six and 12 months on dietary, behavioral, and anthropometric variables were available for 351 and 575 women in the control and intervention groups, respectively (Bhargava and Guthrie, 2002).

Briefly, women in the intervention group, led by nutritionists, met weekly in groups of eight to 15 for the first six weeks, bi-weekly for the next six weeks, and monthly thereafter for nine months. Dietary intakes at baseline, six, and 12 months were measured by a FFQ, and energy and nutrient intakes were calculated. The women's age, marital status and education levels were recorded; household incomes were coded into three groups. Patterns of "mild" and "strenuous" physical exercise for at least 30 minutes were investigated on a scale of 1 to 5 (1 = never, 5 = every day). The women answered, on a scale of 1 to 4 (1 = never, 4 = very often) eight questions regarding their food preferences: whether they (1) liked tasty food; (2) had a craving for rich foods; (3) liked healthy foods; (4) ate more than they should; (5) regarded the food they consumed as satisfying; (6) felt uncomfortable with rich foods; (7) felt deprived in the absence of rich foods; and (8) disliked the taste of fat. The scores on eating healthy foods, feeling uncomfortable with rich foods and disliking the taste of fat were recoded so that higher scores implied less healthy eating. The scores on these eight questions were summed to produce an index of 'unhealthy eating' that ranged from 8 to 32. Last, women's height, weight, waist and hip circumferences were measured at baseline and at six and 12 months.

MODELS FOR WOMEN'S ANTHROPOMETRIC INDICATORS

Body weight responds gradually to nutrient intakes and lifestyle changes and hence weight in the previous time period is an important predictor of current weight. Because of the emphasis in the WHTFSMP on fat intakes, Bhargava and Guthrie (2002) postulated a dynamic random-effects model (Model 1) for the weight (WT) of n subjects using three repeated observations:

$$
\begin{aligned}
\ln (WT)_{it} = a_0 &+ a_1 (Black)_i + a_2 (White)_i + a_3 (Education)_i + a_4 (Income)_i \\
&+ a_5 \ln (Unhealthy\ eating)_i + a_6 (Physical\ exercise)_i + a_7 \ln (Height)_i \\
&+ a_8 \ln (Carbohydrate)_{it} + a_9 \ln (Saturated\ fat)_{it} \\
&+ a_{10} \ln (Monounsaturated\ fat)_{it} \\
&+ a_{11} \ln (Polyunsaturated\ fat)_{it} \\
&+ a_{12} \ln (Energy\ intake)_{it} + a_{13} \ln (WT)_{it-1} \\
&+ u_{1it}\ (i = 1,2,\ldots,n;\ t = 2,3)
\end{aligned}
\tag{7.2}
$$

Here, ln represents the natural logarithm. Coefficients of variables, expressed in logarithms, were the elasticities (percentage change in the dependent variable resulting from a 1% change in the independent variables). The model for weight contained previous measurement as an explanatory variable with coefficient a_{13}. The u_{1it}'s were random-error terms that were also decomposed in a simple random-effects fashion.

Furthermore, a model was developed (Model 2) in which dietary intakes were expressed as ratios to current energy intakes; Model 2 was a special case of Model 1 with the restriction on the coefficients that

$$
a_8 + a_9 + a_{10} + a_{11} + a_{12} = 0
\tag{7.3}
$$

These restrictions were tested using a likelihood-ratio test statistic that was distributed, for large n, as a chi-square variable with 1 degree of freedom.

It was also useful to estimate "static" random-effects models for changes in weight between baseline to six months, and from six to 12 months. Thus, Model 3 was developed for two observations on weight changes that were explained by changes in intakes of carbohydrate, saturated, monounsaturated and polyunsaturated fats, and energy. Alternative variants of Model 3 were estimated for weight changes, with explanatory variables expressed as changes in the *proportions* of energy derived from carbohydrate, and saturated, monounsaturated and polyunsaturated fats (Model 4), and where weight changes were explained by *levels* of these proportions (Model 5).

RESULTS FROM THE ANALYSES OF THE WHTFSMP DATA

The sample means and standard deviations for the control and intervention groups are recorded in Table 7.1. For the control group, there were significant changes between baseline and 12 months in all variables except weight and waist circumference (P < 0.05); changes between six and 12 months were smaller. Mean energy intakes declined from 7223 kJ to 6243 kJ (1 kcal = 4.18 kJ), a 16% reduction. There was a 20% decline in intakes of saturated, monounsaturated and polyunsaturated fats. Even in the absence of counseling, women reduced saturated fat intakes, presumably due to the information (from USDA/DHSS 1990) they received through participating as controls. Changes in the intervention group between baseline and 12 months were statistically significant for all variables (P < 0.05); reduction in energy intake was approximately 27%. Changes in weight, and waist and hip circumferences, were, respectively, 3%, 2% and 1.9%. Intakes of saturated, monounsaturated and polyunsaturated fat were reduced to less than 50% of the baseline levels. There was a 23% increase in energy derived from carbohydrate. Overall, in comparison with the control group, the education program for the intervention group helped to lower fat and energy intakes and promoted weight loss.

The results from estimating models for women's body weights in the control and intervention groups are reported in Table 7.2. Allowing for different means in the two groups, a likelihood ratio test rejected the null hypothesis of parameters constancy in the two groups (chi-square = 56.4, degrees of freedom = 14, P < 0.001). Thus, dynamic models were estimated separately for the control and intervention groups. For the control group, the results for Models 1 and 2 showed that white women and women from higher-income households were significantly lighter (P < 0.05). The index of unhealthy eating was positively associated with weight, while reported frequency of mild physical exercise was negatively associated.

Table 7.1 Sample means and standard deviations of selected variables for the subjects of the Women's Health Trial: Feasibility Study in Minority Populations (WHTFSMP)

	Control group[a]						Intervention group[b]					
	Baseline		6 months		12 months		Baseline		6 months		12 months	
	Mean	SD	Mean	SD	Mean	SD	Mean	SD	Mean	SD	Mean	SD
Age (years)	59.9	6.6					60.1	6.6				
Black (0/1)	0.38	0.49					0.34	0.48				
White (0/1)	0.56	0.50					0.57	0.50				
Education (1–4)	2.91	1.00					2.98	1.00				
Income (1–3)	1.96	0.48					1.98	0.50				
Unhealthy eating (8–32)	21.1	2.84					21.3	2.91				
Physical exercise (1–5)	2.84	1.22					2.86	1.33				
Height (m)	1.62	0.06					1.63	0.06				
Weight (kg)	76.2	12.5	76.0	12.5	75.9	12.7	76.0	12.7	74.0	12.7	73.8	12.8
Waist circumference (m)	0.86	0.11	0.86	0.11	0.86	0.11	0.87	0.11	0.85	0.11	0.85	0.11
Hip circumference (m)	1.09	0.09	1.08	0.09	1.08	0.10	1.09	0.09	1.07	0.09	1.07	0.10
Energy intake (kJ)	7,223	3,262	6,485	3,067	6,243	3,188	7,489	3,736	5,406	2,284	5,448	2,423
Carbohydrate (g)	191	85.6	180	84.0	174	87.3	198	104	181	77.9	186	83.0
Saturated fat (g)	25.2	14.5	21.6	13.0	20.8	14.0	27.1	16.0	13.1	7.7	13.0	8.2
Monounsaturated fat (g)	28.6	15.5	24.3	14.0	22.9	14.3	30.3	16.7	14.0	8.5	13.8	8.6
Polyunsaturated fat (g)	17.0	9.6	14.0	8.4	13.3	9.0	17.5	10.0	7.8	5.3	7.6	4.8
Energy from carbohydrate (%)	45.2	8.0	47.2	8.8	47.5	8.8	44.5	7.7	56.6	9.4	57.7	9.6
Energy from saturated fat (%)	12.8	3.0	12.2	3.0	12.0	3.1	13.3	2.8	8.9	2.6	8.7	2.7
Energy from monounsaturated fat (%)	14.5	3.0	13.8	3.1	13.4	3.1	15.0	2.9	9.6	3.2	9.3	3.2
Energy from polyunsaturated fat (%)	8.7	2.4	8.0	2.5	7.8	2.5	8.8	2.5	5.4	2.1	5.2	2.2

Note: There were 351 subjects in the control group and 575 in the intervention group; dietary intakes were based on food frequency questionnaires.

[a] Changes in all variables except weight and waist circumference between baseline and 12 months were statistically significant (P < 0.05); differences in the changes between baseline and 12 months were significant for all variables except carbohydrate intake.

[b] Changes in all variables between baseline and 12 months were significant; difference in the changes between baseline and 12 months were significant for all variables except carbohydrate intake.

Source: Bhargava and Guthrie (2002).

Table 7.2 Maximum–likelihood estimates of dynamic random-effects models for the weight of women in the WHTFSMP, explained by socio-economic variables and nutrient intakes

	Weight (kg)[a]							
	Control group				Intervention group			
	Model 1		Model 2		Model 1		Model 2	
	Coefficient	SE	Coefficient	SE	Coefficient	SE	Coefficient	SE
Constant	2.100*	0.049	2.197*	0.320	2.094*	0.190	2.114*	0.161
Black (0/1)	0.024	0.017	0.022	0.011	0.032*	0.017	0.032*	0.012
White (0/1)	−0.035*	0.017	−0.038*	0.012	−0.028*	0.014	−0.028*	0.011
Education (1–4)	0.0004	0.003	0.0008	0.002	−0.012*	0.003	−0.012*	0.003
Income (1–3)	−0.008	0.005	−0.010*	0.005	−0.006	0.009	−0.006	0.007
Unhealthy eating (8–32)[b]	0.126*	0.017	0.133*	0.041	0.140*	0.028	0.145*	0.021
Physical exercise, (1–5)	−0.007*	0.003	−0.007*	0.002	−0.009*	0.003	−0.009*	0.002
Height, m	0.776*	0.130	0.804*	0.144	1.061*	0.133	1.074*	0.122
Energy intake, kJ[b]	−0.087*	0.021			−0.028	0.027		
Carbohydrate, g[b]	0.033*	0.013			0.010	0.018		
Saturated fat, g	0.018	0.013			−0.018	0.014		
Monounsaturated fat, g[b]	0.046*	0.020			0.034*	0.017		
Polyunsaturated fat, g[b]	0.0006	0.009			0.007	0.009		
Energy from carbohydrate[b]			0.032*	0.016			0.010	0.013
Energy from saturated fat[b]			0.019	0.016			−0.018	0.012
Energy from monounsaturated fat[b]			0.051*	0.019			0.034*	0.017
Energy from polyunsaturated fat[b]			−0.002	0.010			0.006	0.009
Lagged dependent variable[b]	0.411*	0.029	0.401*	0.088	0.332*	0.045	0.329*	0.047
Between–within variance	8.245*	1.260	8.648*	3.012	9.337*	1.686	9.399*	1.623
Within variance	0.0009		0.0009		0.0011		0.0011	

Note: Values are slope coefficients and standard errors.

[a] Transformed into natural logarithms.

[b] Intakes in Model 2 were expressed as ratios to energy intake.

*P < 0.05.

Source: Bhargava and Guthrie (2002).

These results underscored the importance of behavioral factors, such as craving for rich foods, and physical exercise for body weight. Height was a significant predictor of weight, though the estimated coefficients of height from Models 1 and 2 were significantly lower than the value 2, which would have indicated a preference for combining height and weight as BMI (see equation (7.1)).

The estimated coefficient of current energy intake in Model 1 for the control group was negative and statistically significant. While the likelihood ratio test rejected the restrictions on the coefficients (equation (7.2)) of Model 1 at the 5% significance level, the null hypothesis was accepted at 2.5% level. Moreover, the estimated coefficients of Models 1 and 2 were close, suggesting that conclusions based on either formulation would be similar. The coefficients of monounsaturated fat were positive and significant in Models 1 and 2. This was not true for saturated and polyunsaturated fat intakes, where the estimated coefficients in both models were not statistically different from zero. The estimated coefficients of lagged dependent variables were approximately 0.40 for Models 1 and 2 and were significant, underscoring the need for taking into account the history of weights when investigating the effects of nutrient intakes. Between-to-within variance ratios were large and significant, suggesting that differences between women were partially accounted for by background variables, the index of unhealthy eating, physical exercise patterns, and nutrient intakes.

The results for the intervention group were broadly similar to those for the control group. Black women were heavier and white women were lighter than Hispanic women. The coefficient of education was statistically significant in Models 1 and 2, suggesting that highly educated women benefited more from counseling. Because the coefficient of education was not statistically significant in the control group, awareness of the counseling program was insufficient to induce behavioral changes leading to weight loss. The coefficients of the index of unhealthy eating and of physical exercise patterns in Models 1 and 2 were similar for the control and intervention groups and were statistically significant. However, only intake of monounsaturated fat was positively associated with weight in the intervention group. The coefficient of energy intake in Model 1 was not statistically different from zero. Last, women's age and age-squared were significant predictors of weight.

Table 7.3 presents the results for models for weight changes in the control and intervention groups. In the results from Model 3, the coefficients of energy intakes were estimated with negative signs and were

Table 7.3 Efficient estimates of three versions of static random-effects models for weight changes of the women in the WHTFSMP, explained by changes in carbohydrate, saturated, monounsaturated and polyunsaturated fat, and by transformations of the explanatory variables

	Weight change (kg)											
	Control group						Intervention group					
	Model 3[a]		Model 4[b]		Model 5[c]		Model 3[a]		Model 4[b]		Model 5[c]	
	Coefficient	SE	Coefficient	SE	Coefficient	SE	Coefficient	SE	Coefficient	SE	Coefficient	SE
Constant	−0.001	0.001	−0.001	0.001	0.172*	0.057	−0.006*	0.001	−0.006*	0.001	0.179*	0.044
Change in carbohydrate (g)	0.042*	0.014					0.015	0.011				
Change in saturated fat (g)	0.015	0.011					0.008	0.009				
Change in monounsaturated fat (g)	0.052*	0.014					0.012	0.011				
Change in polyunsaturated fat (g)	−0.003	0.007					0.022*	0.005				
Change in energy (kJ)	0.013*	0.024					−0.054*	0.018				
Change in energy from carbohydrate			0.042*	0.014					0.016	0.011		
Change in energy from Saturated fat			0.016	0.011					0.001	0.009		
Change in energy from Monounsaturated fat			0.052*	0.014					0.012	0.011		
Change in energy from Polyunsaturated fat			−0.004	0.007					0.022*	0.007		
Energy from carbohydrate					0.018	0.011					0.018	0.010
Energy from saturated fat					0.005	0.009					0.006	0.009
Energy from monounsaturated fat					0.024*	0.012					0.010	0.011
Energy from polyunsaturated fat					0.003	0.006					0.016*	0.006

Note: Values are slope coefficients and standard errors; the dependent variable and all explanatory variables were transformed into natural logarithms.

[a] Model 3 explained changes in weight by changes in intakes.

[b] Model 4 explained changes in weight by changes in the percentage of energy derived from particular intakes.

[c] Model 5 explained changes in weight by percentage of energy derived from particular intakes.

*P < 0.05.

Source: Bhargava and Guthrie (2002).

statistically significant for the control and intervention groups. These negative estimates were broadly similar to the estimated coefficients of energy intakes in the model for weight in Table 7.2. Thus, the results again suggested that weight changes might be better explained by changes in the proportions of energy derived from carbohydrate, and saturated, monounsaturated and polyunsaturated fats. The results for Model 4 in Table 7.3 were close to those obtained from estimating Model 3. Moreover, in the control group, the percentages of energy derived from carbohydrate and monounsaturated fat were positively associated with weight changes. By contrast, the percentage of energy derived from polyunsaturated fat was positively associated in the intervention group. Furthermore, the results from Model 5, with explanatory variables in levels, showed positive associations between energy derived from monounsaturated fat and weight changes in the control group. By contrast, a greater proportion of energy from polyunsaturated fat was associated with weight gain in the intervention group. Overall, while the results in Tables 7.2 and 7.3 showed significant effects of macronutrient intakes on body weight and weight changes, the results were ambiguous with respect to possible links between women's fat intakes and body weight. Moreover, the inclusion of protein intakes in Models 1 and 2 led to very similar results and indicated a preference for conversion of dietary intakes as ratios to energy intakes.

In summary, the results from the models for women's weight showed the importance of physical exercise and unhealthy eating habits. Thus, for example, using the question items regarding unhealthy eating habits, investigators can identify women who will face greater difficulties in responding to nutrition education and offer them additional guidance. While the education variable was a significant predictor of body weight in the intervention group, education was not a significant predictor in the control group. Higher education levels may have made it easier for women to understand and apply nutritional guidance provided as part of the intervention. Given the prevalence of obesity among less-educated women, special attention is necessary for devising effective education programs for such women. Last, because high-fat foods are energy-dense, from a practical standpoint a diet with low energy density is likely to be moderately low in fat. The results from WHTFSMP indicate that making lifestyle changes, such as improving dietary habits and increasing physical exercise, can be effective in reducing weight. In view of the high costs of caring for chronic diseases due to obesity, preventive efforts such as nutrition education are sound long-term strategies. Behavioral factors affecting dietary intakes are addressed in section 7.2 and some of the psychological literature on issues of obesity is also discussed.

7.2 Psychological approaches to diet modification and obesity

Food consumption is influenced by individuals' energy and nutrient requirements and by psychological and social factors such as the enjoyment of food in social settings, tastes, dietary knowledge and perceptions of health risks. Moreover, for individuals with poor dietary knowledge, restricting food intakes may be difficult and such individuals can overeat during social interactions or while relaxing. In order to encourage modifications in diets, therefore, detailed knowledge of psychological and cultural factors affecting food consumption is useful. Psychologists and nutritionists have devised several strategies for improving diet quality and lifestyles in developed countries. While the role of economic factors such as food prices is important and is discussed in section 7.3, it is important when designing nutrition education programs, especially for low-income population groups, to focus on the behavioral aspects affecting food consumption.

An important aspect for enhancing dietary knowledge and improving diet quality is that the dietary changes should be sustainable. For example, while individuals may be persuaded via counseling in short-term studies to consume larger quantities of fruits and vegetables and to increase physical activity, it is not feasible to monitor dietary behavior for extended periods. Moreover, even well-motivated individuals may not be able to sustain changes in the long run due to social pressures. For example, if traditional family meals are high in fat, then diets are likely to revert to unhealthy practices that existed prior to nutrition education. It is therefore important to initiate dietary interventions early in life, and numerous school-based interventions have been implemented in countries such as the US (e.g. Reynolds *et al.*, 2000). However, school meals are only a part of overall food intake and if families' food consumption patterns are not healthy, then school-based interventions may not produce long-run improvements. Even so, from a conceptual viewpoint, it is important to understand the types of interventions that are likely to be successful and the time frame over which such interventions might be efficacious.

The psychological literature on dietary change is vast (e.g. Contento, 1995) and cannot be easily condensed into a few paragraphs. However, certain theories, such as the "health belief model" (Rosenstock *et al.*, 1988) and the "stages of change model" (Prochaska and Di Clementi, 1984), have guided much of the research and can be summarized for specific readerships (Glanz and Rimer, 1995). Moreover, as noted previously, economists' definition of rationality is a "substantive" postulate (Simon, 1986) emphasizing the role of prices and incomes for food consumption decisions. In contrast, psychological counseling and increasing dietary knowledge can alter food consumption patterns, so that the "procedural" definition of rationality used by psychologists may be more appealing, especially for populations in developed countries, where food budgets are a relatively small proportion of incomes. Further, psychological research needs to take into account constraints imposed by human physiology, such as individuals' energy and nutritional requirements. For example, heavier individuals have a higher energy requirement and are more likely to consume energy-dense foods, especially if the time available for food consumption is limited. Thus, the efficacy of dietary interventions encouraging lower fat consumption is likely to be enhanced by concomitant programs for weight loss that reduce individuals' energy requirements; an integrated approach to dietary change is likely to be effective. Section 7.2.1 briefly describes the theories underlying psychological research on dietary change. In section 7.2.2, dietary interventions among adult population groups are discussed and the limitations of such interventions are outlined; in section 7.2.3, certain nutrition education programs in school are discussed. Last, in section 7.2.4, the dietary intervention in the WHTFSMP in the US is revisited to illustrate the importance of jointly analyzing psychological, socio-economic and physiological factors affecting food consumption.

7.2.1 Psychological theories guiding interventions for dietary change via behavioral modification

The determinants of human behavior are complex and sustainable changes in diet and lifestyles may be difficult to achieve. While health care professionals can provide similar advice to individuals, individuals' characteristics are likely to determine behavioral changes and compliance. Thus, it would be effective to provide specific advice to individuals based on their behavioral and other characteristics. Psychological theories are useful for understanding the diverse factors affecting human behavior, and integrating theories into counseling programs can enhance dietary changes and the adoption of healthy lifestyles.

A major theme in the "social learning theory" of Bandura (1977) is the constant interaction between individuals and their environments. Such interactions are argued to be bi-directional in that, while individuals are influenced by their environment, individual involvement in turn can change environments. Moreover, knowledge and skills are important for facilitating interactions. For example, better understanding of the risks associated with smoking has not only decreased smoking among well-educated population groups but has also reduced the opportunities for smoking in public settings, due to greater awareness. Furthermore, in social learning theory, individuals' expectations of the results of their actions and their self-efficacy are recognized to be important for making lifestyle changes. Such changes may be motivated by learning from the experience of others, and positive feedback from successful behavioral modifications can encourage further reinforcements. Thus, knowledge of psychological factors is important for designing interventions for dietary and lifestyle changes.

Three sets of models are commonly employed in the psychological literature promoting dietary and behavioral change, i.e. the stages-of-change, health belief, and consumer information-processing models. The stages-of-change model (Prochaska and DiClemente, 1984) defines five stages into which individuals can be placed, i.e. pre-contemplation, contemplation, decision, action, and maintenance. For example, individuals in the pre-contemplation stage have not even thought about making lifestyle changes. In contrast, those in the maintenance stage have not only made necessary changes but are actively maintaining healthy lifestyles. The main advantage in employing the stages-of-change model is that guidance necessary for making changes can be tailored to the stage in which individuals fall. Such models have been useful for the cessation of smoking, partly because it is easy to assess various stages (Prochaska and DiClemente, 1984). By contrast, in the context of dietary change, interpretation of stages can be more difficult; such issues are discussed in section 7.2.2 for individuals in developed countries.

The health belief model was developed by Rosenstock et al. (1988) and emphasizes the role of perceived risks and benefits in making lifestyle changes. Perceived susceptibility to disease and severity are likely to be motivating factors for making dietary and lifestyle changes. In addition, higher perceived benefits can encourage greater changes, while perceived barriers may inhibit or dampen the speed of changes. For example, women may want to lower fat intakes but may not be able to adopt low-fat diets because the high fat content of family meals is a barrier. Moreover, self-efficacy and individuals' readiness for action can facilitate changes given the perception of risks.

The third model commonly used in psychological research, the consumer information-processing model, emphasizes individuals' ability to search for and process information necessary in decision-making. For effective processing of information, it is important to have the information in an appropriate format that can be easily understood. For example, although detailed labels describing food contents are useful for consumers in making choices, presenting too much information can obscure the salient aspects. Moreover, the information-processing model recognizes that individuals will devise rules of thumb in decision-making and that current choices will influence subsequent decisions. This is also in the spirit of models of habit formation used in the economics literature, which are motivated by the reasoning that "choices depend on tastes, and tastes depend on past choices" (Gorman, 1968).

Several features of the models used in psychological applications are useful for promoting dietary and lifestyle changes. However, the relevance of the postulates depends on the settings. Moreover, to encourage dietary changes, it would be helpful to devise specific advice depending on individuals' characteristics and their understanding of the perceived risks of over-consumption of food. For analyzing the determinants of dietary and lifestyle changes, it is desirable to measure individual characteristics such as dietary knowledge and awareness levels, motivation, self-efficacy, attention to nutritional labels, perception of health risks, etc. While some studies have attempted to measure several aspects, most studies are driven by specific hypotheses with a somewhat narrow focus. For example, stages-of-change models have been extensively employed in dietary interventions for reducing fat intakes. However, as discussed below, individuals with low education levels may have difficulties in correctly estimating their fat intakes and hence the analyses may be subject to misinterpretation.

7.2.2 Nutrition education programs and dietary change among adults in developed countries

There have been several nutrition education interventions in the US to improve dietary intakes and lifestyles. The study designs have been guided by psychological theories and epidemiological evidence on the effects of excessive nutrient intakes on health outcomes. For example, the Women's Health Trial was a randomized controlled trial that attempted to decrease fat intakes by white women living in Seattle, Houston and Cincinnati so as to reduce their risks of breast cancer (Henderson *et al.*, 1990). This trial was followed by the WHTFSMP, discussed in section 7.1.4 (Bowen *et al.*, 1996). The ongoing Women's Health Initiative is the largest randomized controlled trial, with over 48,000 women participating (Prentice *et al.*, 2006). Behavioral questionnaires used in these trials covered many aspects of stages-of-change and health belief models; the proximate determinants of dietary changes will be discussed in section 7.2.4 using data from WHTFSMP. However, it will be useful to describe some related studies, in order to gain a broader understanding of strategies used to improve dietary intakes and lifestyles in developed countries.

The stages-of-change models have been extensively used in dietary research for reducing fat intakes. Individuals in these studies are classified into the five stages (pre-contemplation, contemplation, decision, action and maintenance), based on their responses to several items in the questionnaires (Glanz *et al.*, 1994). In promoting low-fat diets, for example, questions such as "How high is fat in your overall diet?", "For how long have you followed a diet that is low in fat?", "How high is fiber in your diet?", and "For how long have you followed a diet that is high in fiber?" are typically answered on a scale from 1 to 5. Moreover, behavioral intentions to change one's diet in the next few months, and changes in eating habits and success in reducing fat intakes and increasing fiber intakes, are investigated. Glanz *et al.* (1994) used the data from the Working Well Study to investigate the percentages of intakes of energy from fat by approximately 17,000 individuals. Percentages of energy derived from fat for individuals in the five stages were 39.6, 39.3, 39.7, 37.4 and 31.7, respectively. Percentages of energy from fat were therefore significantly lower ($P < 0.05$) in the action and maintenance stages. Similar results were obtained in other studies that attempted to lower fat intakes (e.g. Ounpuu *et al.*, 2000). However, as noted in section 7.1.1, assessment of dietary intakes via FFQs and other methods is subject to internal ("within-subject") variation in the intakes, thereby widening confidence intervals for mean intakes. In practice, therefore, one might see significantly lower fat intakes, mainly by individuals in the maintenance stage.

Furthermore, stages-of-change models are useful for designing different strategies for individuals in the five stages. For example, moving individuals from the pre-contemplative stage to later stages may involve cognitive processes such as counseling individuals on the adverse consequences of diets high in fat. By contrast, behavioral processes such as advice on the preparation of low-fat meals may be more important for individuals in the action and maintenance stages (Ounpuu *et al.*, 2000). Even so, the five stages are broad categories and one is likely to observe statistically significant differences, mainly for individuals in the action and maintenance stages. Moreover, items assessing individuals' characteristics, such as their attitudes towards health, self-efficacy and other aspects, can provide guidance for refining nutrition education programs. For example, certain individuals may be concerned about specific medical conditions and may be willing to make appropriate changes that can be facilitated by higher education levels. Smith *et al.* (1995) found that in a sample of 191 Australians, individuals with "belief in diet–heart disease [relationship]", "[with high] initial food-guide based knowledge", and "motivation [to reduce] plasma cholesterol" were significantly more likely to make dietary changes. Steptoe *et al.* (2000) reported that motivating factors such as "behavioral intention" and "anticipation regret [for not making dietary changes]" were significant predictors of lower fat intakes among 681 patients in the UK with heart disease and/or high LDL cholesterol concentrations.

Another important aspect of nutrition education programs is to assess their long-term effects. In the Women's Health Trial, for example, follow-up after a year showed that reductions in fat intakes achieved during the trial were maintained, though at a lower level (Urban *et al.*, 1992). Because of the high costs of counseling and monitoring individuals in developed countries, studies seldom extend beyond one or two years; the case for interventions is stronger if health benefits endure over time. While lasting changes may be difficult to achieve among the general population, patients that have suffered treatments against cancer and coronary heart disease may be more willing to adopt a healthy diet and lifestyle. However, well-educated population subgroups can make lasting behavioral changes, provided that they correctly perceive risks and have a high degree of self-efficacy. Even for individuals with low education levels, counseling through community and religious groups such as churches has increased fruit and vegetable intakes (Resnicow *et al.*, 2001). However, as noted in section 7.1, if individuals are severely overweight or obese, then their energy requirements are high and they are more likely to consume energy-dense foods, thereby making dietary changes more difficult. Most interventions incorporating psychological aspects have not addressed physiological factors, such as the higher energy needs of overweight individuals. These issues are discussed in section 7.2.4, using the data from the WHTFSMP, after the description of some interventions for schoolchildren in the US.

7.2.3 *Nutrition education interventions for schoolchildren*

Obesity among children is on the rise in developed countries and intervention programs can target children, parents, teachers and cafeterias to promote healthful eating and lifestyles. Certain psychological factors for adults, such as risk perceptions, also apply to children. However, food consumption patterns at home and the dietary habits of peers can significantly affect children's intakes. For example, eating out is a common form of socialization among adolescents, and income constraints may force children from poor backgrounds to frequent fast-food restaurants. While dietary modifications can be achieved by incorporating psychological constructs for adults (e.g. Baranowski *et al.*, 2002), it is important to involve parents, teachers and cafeteria staff to improve diet quality and maintenance of healthy body weight. Certain studies in the US, UK and Greece are discussed next, because early interventions can circumvent or delay the development of chronic diseases associated with obesity.

A comprehensive nutrition education program was implemented for approximately 1,700 US families with children in the third grade (age 8–9) in 1994 to increase the consumption of fruits and vegetables among both children and parents (Reynolds *et al.*, 2000). The study involved 28 schools and half the schools received nutrition education along the lines of the "five a day program" (National Cancer Institute, 2001). The intervention had three components: in the classroom, for parents at home, and for cafeteria staff. Children and their parents were observed at baseline and at two follow-ups at one and two years, respectively, after the baseline. Methods for increasing knowledge in the classroom were devised by psychologists and included components such as self-monitoring, problem-solving, reinforcement and taste-testing. The parent component encouraged parents and children to modify their dietary behaviors. The food services component taught cafeteria staff to purchase and prepare foods with higher quantities of fruits and vegetables. Children's intakes were measured via the 24-hour recall method. Psychosocial aspects such as "asking skills" and self-efficacy were assessed for children. Parents also completed psychosocial questionnaires covering aspects such as the perceived health benefits of nutrition education.

The main findings of this "High 5" study were that the fruit and vegetable intakes of children in the intervention groups increased significantly more than those in the control group from baseline to first follow-up, and also from baseline to second follow-up. However, there was a drop in fruit and vegetable consumption between the first and second follow-ups.

Percentages of energy derived from total and saturated fats were significantly lowered by the intervention. Moreover, in the intervention group there was an increase in energy derived from carbohydrates, and intakes of desirable nutrients such as fiber, β-carotene, folate and vitamin C also increased significantly in the intervention group. However, for parents, increases in fruit and vegetable consumption between baseline and first follow-up were significant only for white families. Overall, nutrition education intervention was successful in increasing fruit and vegetable intakes, which are important for lowering the risks of chronic diseases such as cancers. However, plasma levels of nutrients were not measured in this study, which could have provided further evidence on compliance and on the long-run benefits of the intervention.

Another study in the US looked at 787 children enrolled in the "Head Start" program and investigated the effects of food service intervention and supplemental nutrition education intervention on *total* plasma cholesterol levels (Williams *et al.*, 2004). The intervention reduced total plasma cholesterol levels by approximately 6 mg/dl in comparison with the control group. While the intervention seemed quite effective, total plasma cholesterol was not disaggregated into LDL and HDL components. In fact, a decrease in LDL cholesterol is beneficial for reducing the risks of cardiovascular disease, while *higher* HDL cholesterol is associated with lower risks of heart disease (Bhargava, 2003b). Thus, in the absence of data on these cholesterol subgroups, it is difficult to assert that the intervention was successful in lowering the risks of cardiovascular disease. However, it should be noted that a study of 1,510 students enrolled in the first grade in Greece found that LDL cholesterol was significantly lowered by a nutrition education program, though HDL cholesterol did not decrease significantly. Also, it is known from Keys (1984) that LDL cholesterol is more responsive to dietary changes. Thus, the decline in total cholesterol in the intervention group in the study of Williams *et al.* (2004) was probably due to a significant decrease in LDL cholesterol. However, it is essential to have data on both LDL and HDL cholesterol in order to reach definite conclusions, and most studies disaggregate the total cholesterol levels.

Last, the effects of nutrition education programs in schools have been found to be ambiguous, depending on settings (Campbell *et al.*, 2001). In a randomized trial involving 734 children in the age group 7–11 years, Sahota *et al.* (2001) examined the effects of a nutrition education intervention on children's diets, BMI, physical activity and psychological measures. The authors concluded that only fruit and vegetable intakes were significantly increased by the intervention. In general, while nutrition education and physical activity programs in schools are likely to affect the short-term diets and behaviors of children, such changes may not be sustainable in the long run. Moreover, it is expensive to monitor children over extended periods, and it would not seem cost-effective for studies to span several years, especially if the benefits from lifestyle changes appear to be small. In contrast, improving diet quality in the cafeteria and more strenuous physical activity programs for all schoolchildren may afford greater health benefits. The marginal costs of extending existing school programs are likely to be low and the benefits of increased physical activity can be substantial in the long run, especially in view of the increase in child obesity in recent years.

7.2.4 Proximate determinants of dietary intakes in the Women's Health Trial: Feasibility Study in Minority Populations

The WHTFSMP data set was described above in section 7.1.4. The questionnaire also included several behavioral items reflecting theories pertaining to eating habits, food preparation time and budget, health beliefs and motivation (Bhargava and Hays, 2004). Thus, in addition to the "unhealthy eating" index described in section 7.1.4, an index called "preparation and budget" was based on answers, on a scale of 1 to 4 (1 = very little, 4 = a lot), to four questions regarding time spent shopping for food, preparing food, trouble taken in preparing food and money spent on food. The scores on these four questions were summed and this index ranged from 4 to 16. Another index, "Concerned about health", reflecting the health belief model, was based on answers to seven questions, on a scale of 1 to 5, regarding health-related reasons for participating in the WHTFSMP. This index summed the scores on whether the subjects (1) had a relative or a friend with cancer; (2) were trying to lose or control weight; (3) were concerned about their health; (4) were concerned about their husband's health, (5) were concerned about their family's health; (6) had high cholesterol or heart problems; and (7) were afraid of getting cancer. Next, an index, "Participation motivation", was based on answers to questions regarding motivating factors for participating in the WHTFSMP. The women answered four questions, on a scale of 1 to 5, about whether they felt that their participation was important for helping scientific research, learning more about nutrition, sharing activities with new people, and helping the community. In addition to these four indices, women were asked if they were consuming a low-fat diet (0 = no, 1 = yes), and on a scale of 1 to 4 (1 = not strong, 4 = very strong), about their 'Desirability of change' to low-fat diet, reflecting self-efficacy.

Bhargava and Hays (2004) postulated dynamic random-effects models for dietary intakes by n women at three time points (baseline, six and 12 months). The model was estimated for intakes of carbohydrate, saturated, monounsaturated and polyunsaturated fats, fiber, β-carotene, ascorbic acid and calcium. The models for dietary intakes contained intakes in the previous period as explanatory variables and the results for the control and intervention groups are discussed in the test of this subsection.

EMPIRICAL RESULTS FOR DIETARY INTAKES IN THE CONTROL GROUP

The empirical results from dynamic random-effects models for intakes by women in the control group are reported in Table 7.4. The models for dietary intakes were estimated separately for the control and intervention groups because likelihood ratio tests rejected the null hypothesis that parameters were constant in the two groups. Another set of likelihood ratio statistics were applied to test the null hypotheses that dietary intakes should be expressed as ratios to energy intake. The results indicated that it was preferable to adjust for energy intakes by including energy intake as an explanatory variable in the models. First, for intakes of carbohydrate and of saturated, monounsaturated and polyunsaturated fats, the indicator variables for race, education and income were not significant predictors. Height was a significant predictor of these intakes, though weight was not significant in the models for carbohydrate and polyunsaturated fat intakes. Both height and weight were significant predictors of intakes of saturated and monounsaturated fats. Application of a likelihood ratio test rejected the null hypothesis that height and weight should be combined as BMI in the models for dietary intakes.

The unhealthy eating index was positively associated with intakes of saturated and monounsaturated fats but not with intakes of carbohydrate and polyunsaturated fat. The food preparation and budget index was positively associated with monounsaturated and polyunsaturated fat intakes. The indicator variable for whether the woman was eating a low-fat diet was positively associated with carbohydrate intake and negatively associated with fat intakes. The categorical variable "Desirability of change to low-fat diet" was negatively associated with saturated and monounsaturated fat intakes. The coefficient of the concerned about health index was significant in the models for monounsaturated and

Table 7.4 Maximum-likelihood estimates of dynamic random-effects models for intakes of carbohydrate, saturated, monounsaturated and polyunsaturated fats, fiber, β-carotene, ascorbic acid and calcium by women in the WHTFSMP control group, explained by anthropometric, socio-economic and behavioral variables

	Intakes[a]							
	Carbohydrate (g)	Saturated fat (g)	Monounsaturated fat (g)	Polyunsaturated fat (g)	Fiber (g)	β-carotene (μg)	Ascorbic acid (mg)	Calcium (mg)
Constant	−1.775* ± 0.099	−5.740* ± 0.393	−5.250* ± 0.045	−4.986* ± 0.172	−3.121* ± 0.062	1.366 ± 0.972	−0.938* ± 0.038	−1.884* ± 0.505
Black (0/1)	−0.063 ± 0.067	0.058 ± 0.084	0.094 ± 0.085	0.114 ± 0.107	−0.196* ± 0.090	0.129 ± 0.210	−0.001 ± 0.157	−0.139 ± 0.122
White (0/1)	−0.081 ± 0.067	0.064 ± 0.083	0.033 ± 0.084	0.053 ± 0.015	−0.089 ± 0.088	0.179 ± 0.207	−0.089 ± 0.157	0.116 ± 0.124
Education (1–4)	0.001 ± 0.009	0.005 ± 0.011	−0.001 ± 0.011	−0.008 ± 0.014	0.022 ± 0.015	0.056* ± 0.027	0.040* ± 0.021	0.002 ± 0.016
Income (1–3)	−0.001 ± 0.018	−0.037 ± 0.024	−0.008 ± 0.022	0.019 ± 0.029	0.052 ± 0.032	−0.105 ± 0.057	−0.001 ± 0.043	0.013 ± 0.035
Height (m)[a]	0.521* ± 0.126	−0.841* ± 0.192	−0.883* ± 0.148	−0.706* ± 0.276	0.617* ± 0.037	0.733 ± 0.395	1.486* ± 0.139	0.114 ± 0.312
Unhealthy eating (8–32)[a]	0.030 ± 0.019	0.157* ± 0.054	0.088* ± 0.021	−0.056 ± 0.049	−0.034 ± 0.055	0.003 ± 0.127	−0.060 ± 0.051	−0.008 ± 0.033
Preparation and budget (4–16)[a]	−0.058 ± 0.032	−0.011 ± 0.026	0.071* ± 0.023	0.088* ± 0.023	−0.042 ± 0.052	−0.089 ± 0.069	0.052* ± 0.021	−0.023 ± 0.061
Low-fat diet (0/1)	0.064* ± 0.014	−0.092* ± 0.019	−0.104* ± 0.018	−0.121* ± 0.023	0.126* ± 0.024	0.240* ± 0.042	0.200* ± 0.035	0.079* ± 0.027
Desirability of change (1–4)	0.014 ± 0.008	−0.033* ± 0.011	−0.032* ± 0.010	−0.024 ± 0.014	0.017 ± 0.014	0.002 ± 0.024	0.026 ± 0.020	0.039* ± 0.015
Concerned about health (7–35)[a]	−0.019 ± 0.020	0.017 ± 0.032	0.047* ± 0.017	0.052* ± 0.023	−0.105* ± 0.017	−0.049 ± 0.080	−0.180* ± 0.041	−0.061* ± 0.030
Participation motivation (4–20)[a]	0.026 ± 0.017	−0.044 ± 0.037	−0.062 ± 0.033	−0.080* ± 0.037	0.054 ± 0.033	0.123 ± 0.085	0.122* ± 0.046	0.102* ± 0.042
Weight (kg)[a]	−0.014 ± 0.048	0.135* ± 0.046	0.083* ± 0.041	0.046 ± 0.039	0.021* ± 0.004	0.148 ± 0.147	−0.107* ± 0.012	0.121 ± 0.083
Energy intake (kJ)[a]	0.922* ± 0.014	1.139* ± 0.017	1.108* ± 0.015	1.038* ± 0.025	0.733* ± 0.025	0.715* ± 0.042	0.602* ± 0.037	1.005* ± 0.021
Lagged dependent variable[a]	0.003 ± 0.017	−0.002 ± 0.017	0.011 ± 0.016	0.059* ± 0.022	0.034 ± 0.036	0.029 ± 0.055	0.177* ± 0.049	0.012 ± 0.021
Between–Within variance	0.985* ± 0.143	1.051* ± 0.147	0.820* ± 0.120	0.730* ± 0.118	1.274* ± 0.230	1.223* ± 0.282	0.720* ± 0.181	1.089* ± 0.159
Within variance	0.0150	0.0254	0.0259	0.0459	0.0411	0.1262	0.1052	0.0511
Chi-square (3)[3]	1.77	6.57	3.31	0.35	0.59	9.51*	1.43	0.002

Note: Values are slope coefficients ± standard errors.

[a] Transformed into natural logarithms.

[b] Chi-square test for exogeneity of energy intake (3 degrees of freedom).

*P < 0.05.

Source: Bhargava and Hays (2004).

polyunsaturated fats, with *positive* signs implying that women in the control group who had greater concerns for health were consuming higher quantities of these two fats. The participation motivation index was significant only for intakes of polyunsaturated fat. The coefficients of energy intakes were large and statistically significant. The coefficients of previous intakes were small and statistically insignificant in the models for carbohydrate and saturated and monounsaturated fat intakes, but was significant for polyunsaturated fat intakes. Chi-square tests accepted the null hypothesis that the random effects were uncorrelated with energy intakes.

The results from the models for intakes of fiber, β-carotene, ascorbic acid, and calcium showed that black women in the control group consumed significantly lower quantities of fiber. Better-educated women consumed greater quantities of β-carotene and ascorbic acid. The food preparation and budget index was positively associated only with ascorbic acid intakes. Women who reported consuming low-fat diets had higher intakes of fiber, β-carotene, ascorbic acid and calcium. Women's height and weight were significant predictors of intakes of fiber and ascorbic acid. Women reporting greater concerns about health consumed significantly lower quantities of fiber, ascorbic acid, and calcium. Moreover, women who scored higher on the participation motivation index consumed greater amounts of ascorbic acid and calcium. The coefficients of energy intakes were large and statistically significant in models for fiber, β-carotene, ascorbic acid and calcium. The coefficients of previous intakes were small and significant only in the model for ascorbic acid intakes; between–within variance ratios were large and statistically significant.

EMPIRICAL RESULTS FOR THE DIETARY INTAKES IN THE INTERVENTION GROUP

The empirical results for dietary intakes by women in the intervention group are shown in Table 7.5. Intakes of monounsaturated and polyunsaturated fats were significantly higher among black and white women. Better-educated women had significantly lower intakes of saturated, monounsaturated and polyunsaturated fats. This was in contrast with the results in Table 7.4, where education was not a significant predictor of fat intakes in the control group. The unhealthy eating index was negatively associated with carbohydrate intakes and positively associated with fat intakes, with statistically significant coefficients. The coefficients of the indicator variable for eating a low-fat diet were statistically significant for carbohydrate and fat intakes. The desirability of change variable was

Table 7.5 Maximum-likelihood estimates of dynamic random-effects models for intakes of carbohydrate, saturated, monounsaturated and polyunsaturated fats, fiber, β-carotene, ascorbic acid, and calcium by women in the WHTFSMP intervention group, explained by anthropometric, socio-economic and behavioral variables

	Intakes[a]							
	Carbohydrate (g)	Saturated fat (g)	Monounsaturated fat (g)	Polyunsaturated fat (g)	Fiber (g)	β-carotene (μg)	Ascorbic acid (mg)	Calcium (mg)
Constant	−1.525* ± 0.167	−5.419* ± 0.027	−5.310* ± 0.220	−5.610* ± 0.054	−4.053* ± 0.300	0.766 ± 0.534	−1.386* ± 0.461	−1.484* ± 0.166
Black (0/1)	−0.038 ± 0.028	0.054 ± 0.040	0.208* ± 0.044	0.238* ± 0.043	−0.049 ± 0.048	−0.001 ± 0.080	−0.012 ± 0.070	−0.273* ± 0.053
White (0/1)	−0.052* ± 0.026	0.005 ± 0.038	0.092* ± 0.043	0.131* ± 0.043	0.056 ± 0.046	0.032 ± 0.077	−0.068 ± 0.068	−0.056 ± 0.051
Education (1–4)	0.005 ± 0.006	−0.029* ± 0.009	−0.040* ± 0.010	−0.026* ± 0.011	0.039* ± 0.011	0.042* ± 0.019	0.052* ± 0.016	0.013 ± 0.013
Income (1–3)	0.009 ± 0.013	−0.012 ± 0.014	−0.032 ± 0.021	−0.029 ± 0.020	0.024 ± 0.022	0.047 ± 0.038	0.027 ± 0.033	0.047 ± 0.026
Height (m)[a]	0.180 ± 0.135	−0.612* ± 0.066	−0.761* ± 0.115	−1.003* ± 0.071	0.049 ± 0.051	−0.022 ± 0.323	0.464* ± 0.229	0.728* ± 0.140
Unhealthy eating (8–32)[a]	−0.135* ± 0.022	0.314* ± 0.033	0.345* ± 0.041	0.245* ± 0.025	−0.209* ± 0.053	−0.176 ± 0.092	−0.306* ± 0.087	−0.101* ± 0.038
Preparation and budget (4–16)[a]	−0.009 ± 0.024	0.001 ± 0.025	0.006 ± 0.040	0.015 ± 0.021	−0.016 ± 0.032	−0.086 ± 0.081	−0.062 ± 0.041	0.057 ± 0.050
Low-fat diet (0–1)[a]	0.134* ± 0.018	−0.212* ± 0.028	−0.237* ± 0.031	−0.219* ± 0.035	0.179* ± 0.033	0.100 ± 0.058	0.190* ± 0.045	0.170* ± 0.034
Desirability of change (1–4)	0.020* ± 0.007	−0.035* ± 0.008	−0.051* ± 0.012	−0.048* ± 0.010	0.030* ± 0.013	0.053* ± 0.023	0.021 ± 0.016	0.026 ± 0.014
Concerned about health (7–35)[a]	−0.021 ± 0.014	−0.062* ± 0.013	−0.057* ± 0.019	−0.006 ± 0.022	0.060* ± 0.030	0.203* ± 0.033	0.171* ± 0.021	0.040 ± 0.036
Participation motivation (4–20)[a]	0.047* ± 0.022	−0.080* ± 0.026	−0.128* ± 0.035	−0.136* ± 0.047	0.075* ± 0.036	0.179* ± 0.018	0.052 ± 0.037	−0.003 ± 0.041
Weight (kg)[a]	−0.040* ± 0.019	0.015* ± 0.007	0.075 ± 0.049	0.134* ± 0.008	0.083* ± 0.020	0.090* ± 0.013	0.036 ± 0.039	−0.080 ± 0.070
Energy intake (kJ)[a]	0.965* ± 0.004	1.109* ± 0.005	1.082* ± 0.010	1.027* ± 0.007	0.823* ± 0.013	0.674* ± 0.039	0.674* ± 0.032	1.044* ± 0.009
Lagged dependent variable[a]	0.014 ± 0.008	0.040* ± 0.012	0.045* ± 0.012	0.047* ± 0.014	0.044 ± 0.023	0.144* ± 0.044	0.140* ± 0.037	0.032* ± 0.006
Between-within variance	0.784* ± 0.089	0.509* ± 0.069	0.590* ± 0.076	0.594* ± 0.077	0.729* ± 0.105	0.486* ± 0.110	0.793* ± 0.133	0.877* ± 0.090
Within variance	0.0150	0.0440	0.0496	0.0655	0.0517	0.1805	0.0995	0.0575
Chi-square (3)[3]	8.82*	6.02	4.39	4.07	1.10	15.58**	1.15	2.12

Note: Values are slope coefficients ± standard errors.

[a] Transformed into natural logarithms.

[b] Chi-square test for exogeneity of energy intake (3 degrees of freedom).

*P < 0.05.

Source: Bhargava and Hays (2004).

significant and positively associated with carbohydrate intakes and negatively associated with fat intakes. The concerned about health index was estimated with *negative* coefficients that were significant in the models for saturated and monounsaturated fat intakes. Thus, in contrast with control group, women in the intervention group who were more concerned about health had lower intakes of monounsaturated and polyunsaturated fats. The participation motivation index was significantly and positively associated with carbohydrate intakes and negatively associated with fat intakes. Intakes in the previous period were estimated with small coefficients that were significant in the models for saturated, monounsaturated and polyunsaturated fat intakes.

Furthermore, the results in Table 7.5 show that better-educated women consumed higher quantities of fiber, β-carotene and ascorbic acid. Calcium intakes were significantly lower for black women, presumably due to lower consumption of dairy products. Women with higher scores on the unhealthy eating index consumed significantly lower quantities of fiber, ascorbic acid and calcium. Women consuming low-fat diets had higher intakes of fiber, ascorbic acid and calcium. The coefficient of this indicator variable in the model for β-carotene was not statistically significant, presumably due to large within-subject variation in these intakes. The desirability of change variable was significantly associated with intakes of fiber and β-carotene. Women with higher scores on the concerned about health index had significantly higher intakes of fiber, β-carotene and ascorbic acid. The coefficients of the participation motivation index were significant for intakes of fiber and β-carotene in the intervention group. Previous intakes were significant in the models for β-carotene, ascorbic acid and calcium.

SOME IMPLICATIONS OF THE FINDINGS FROM THE WHTFSMP

First, it should be noted that the coefficients of the indicator variable for whether the women were eating a low-fat diet were significant for fat intakes in the control and intervention groups in Tables 7.4 and 7.5, respectively. However, in the control group, for subjects reporting eating a low-fat diet, the average percentages of energy derived from fat at baseline and 12 months were, respectively, 33.6 and 31.1. Corresponding averages for the intervention group were 34.5 and 22.5, respectively. Thus, at baseline, women in the two groups had similar perceptions regarding what constituted a low-fat diet. In fact, at 12 months, the average subject in the intervention group was consuming a low-fat diet.

It would therefore seem important to refine questions that are posed to investigate stages of change in behavioral research (e.g. Glanz, 1997). For example, it would be useful to investigate individuals' understanding of foods that are low in fat prior to classifying them into five stages using the algorithms mentioned in section 7.2.2.

Second, the concerned about health index was positively associated with intakes of monounsaturated and polyunsaturated fats and negatively associated with fiber, ascorbic acid and calcium intakes in the models for the control group. A possible explanation for these findings is that women more concerned about health were in fact those consuming higher quantities of monounsaturated and polyunsaturated fats and lower quantities of fiber, ascorbic acid and calcium. By contrast, in the intervention group, the concerned about health index was *negatively* associated with intakes of saturated and monounsaturated fats and was *positively* associated with fiber, β-carotene and ascorbic acid intakes. Thus, women in the intervention group who perceived higher health risks made appropriate dietary changes. Because education in the control group was not a significant predictor of intakes, nutrition education programs would seem essential for encouraging dietary and lifestyle changes among women in the US.

Finally, the results underscored the role of behavioral factors on dietary intakes by minority women and women of low socio-economic status. While the importance of such factors has been recognized in nutrition education research (Contento, 1995) and by policy-makers (USDA, 1999; National Cancer Institute, 2001), there have been few studies facilitating the development of more targeted approaches to nutrition education. Results from the analysis of the WHTFSMP data showed the importance for improving dietary intakes of women's perceptions regarding health and of their motivation for change. These factors can be investigated via brief questionnaires given to individuals prior to the intervention program. Nutrition education programs incorporating responses are likely to be more effective in facilitating dietary change. Moreover, knowledge of cultural barriers to dietary change such as composition of family meals, consumed especially at weekends (Bhargava *et al.*, 1994), can provide additional insights. The design of efficacious nutrition education programs will benefit from such considerations, though implementing the programs is likely to be expensive. However, the costs of nutrition education programs should be compared with the costs of treatments for chronic conditions that are associated with obesity, such as diabetes, hypertension and cardiovascular disease. Overall, prevention is a cost-effective approach and requires close collaboration between dietitians, nutritionists, physicians and social scientists.

7.3 Approaches to obesity in economics

Due to recent increases in obesity rates in developed and developing countries (Popkin and Doak, 1998), economists are interested in analyzing the determinants of body weights and in devising policies for stemming the epidemic. Economic factors such as low food prices and high incomes can promote over-consumption of food. Moreover, as the costs associated with the treatment of obesity-related diseases such as hypertension, diabetes and coronary heart disease are large, health economists are naturally concerned with obesity (e.g. Finkelstein *et al.*, 2004). As noted in the preceding chapters, the determinants of body weights are complex, and the effects of food intakes on weight are analyzed in nutrition research (section 7.1), while individuals' desire to make dietary changes are the focus of psychological studies (section 7.2). For the economist, it is important to analyze the effects of food prices and incomes on food consumption, which in turn may lead to weight gain. However, certain conceptual issues can limit the scope of economic analyses. For example, it may be difficult to claim that the wide availability of inexpensive and palatable foods is the "cause" of weight gain, as individuals can limit their intakes. Historically, food was scarce and expensive, and individuals had to expend considerable energy obtaining food; obesity rates were also low in such circumstances. In a sense, while economic variables critically affect food availability and prices, economic analyses can mainly investigate trends underlying food consumption. In section 7.3.1, some difficulties in analyzing food consumption data in developed countries are briefly outlined, where numerous foods are available. Section 7.3.2 discusses some economic studies investigating the effects on body weights of factors such as improvements in food production technologies, access to restaurants and the opportunity cost of time spent on cooking. Section 7.4 summarizes the implications of nutritional, psychological and economic approaches for reducing the prevalence of obesity.

7.3.1 *Modeling the effects of food prices and incomes on food consumption in developed countries*

Demand analyses using data on households in developed countries are known to be complex from the standpoint of implementing theoretical economics models (e.g. Prais and Houthakkar, 1955; Deaton and Muellbauer, 1980). Moreover, modeling the demand for food is complicated by the large number of foods available in grocery stores and by the degree of substitutability between various food subgroups. For example, the Pyramid Serving Database for the US (National Cancer Institute, 2006) consists of 4,542 food items which are used to convert food intakes into energy and nutrient intakes. While it is possible to aggregate foods in broad categories, even in elaborate household surveys it is difficult to record the prices of all food items that households purchase. For example, the National Food Stamp Program Survey (NFSPS) in the US (Cohen *et al.*, 1999) recorded prices of up to 50 foods purchased during the week by households. Thus, it is appealing to estimate characteristics models for demand as proposed by Gorman (1980), where food characteristics such as energy, protein, vitamin and mineral content are factors driving demand. Such models can be useful in economic applications, especially since the number of food groups is large, while the number of characteristics is likely to be less than (say) 20 nutrients.

In an analysis of the data from the National Food Survey in the UK, Pudney (1981) tested the implications of the characteristics model for demand for 19 food groups (beef and veal, mutton and lamb, pork, corned meat, uncooked bacon and ham, cooked bacon and ham, other cooked meat, other canned meat, liver, offals (excluding liver), poultry, pork sausages, beef sausages, other meat products, fresh white fish, processed white fish, processed fat fish, cooked fish, and canned and bottled fish). Some characteristics affecting demand were postulated to be "bulk", protein, caloric value and preparation time. Restrictions on model parameters implied by the characteristics model for demand were tested using chi-square statistics based on instrumental variables estimation. Characteristics models are in the spirit of the latent variables framework (Joreskorg, 1970), since characteristics reflect unobserved variables underlying the demand. Pudney (1981: 433) noted that while most models for demand for food performed poorly using real data, the characteristics model provided some evidence that considerations such as the protein content of foods were important for explaining demand. However, simplifying assumptions invoked by Pudney (1981) complicated the interpretation of the results. This is because individuals' food consumption patterns evolve gradually over time, and nutrients such as vitamins A and C are important characteristics on which individuals base their decisions when purchasing fruits and vegetables. However, the analysis by Pudney (1981) ruled out these characteristics *a priori*. In fact, shoppers typically buy a mix of meat, milk, fruits and vegetables, and carbohydrates, and the proportions in which such foods are consumed determine the quality of diet in developed countries. In the analysis of the WHTFSMP data in section 7.1.3, the models for body weights indicated a preference for transforming nutrient intakes as ratios to energy intakes, which is consistent with mixing of different food groups in the diet. Thus, while the characteristics model of demand for food is an appealing formulation, further research is necessary in order to make its assumptions consistent with evidence from the nutritional sciences.

To investigate the effects of prices on food consumed ("used"), Yen *et al.* (2003) used trans-logarithmic demand functions and data from the NFSPS to estimate the price and income elasticities of 13 food groups (dairy products, fats and oils, cereals, bread, beef, pork, poultry, fish, other meat, eggs, vegetables and fruits, juices and soft drinks, and mixed foods). The main advantage in employing trans-logarithmic functions is flexibility with respect to the multiplicative effects of explanatory variables. However, despite the aggregation of food subgroups, zero consumption was reported for many groups due to the large variety of foods consumed; the authors' estimation methods took into account the zero values. The estimated elasticities can be used for policy analysis in that one might subsidize (or tax) certain food groups to encourage (or discourage) consumption. However, a potential drawback in estimating trans-logarithmic demand functions is the large numbers of parameters in the systems, which makes it difficult to interpret the coefficients. Also, the magnitudes of elasticities and their standard errors depend on the number of food groups selected for the analysis. The NFSPS was carried out in 35 Primary Sampling Units (PSUs) and it was possible to construct average prices for the PSU using data on households. However, Yen *et al.* used "regional" prices, which entailed a much greater degree of aggregation. Thus, it would seem necessary to conduct further analyses using NFSPS data to analyze the effects of food prices on household consumption. Moreover, from a policy standpoint, it is necessary to identify specific foods that should be subsidized or taxed given consumption patterns. For example, while taxing "junk" foods may seem a useful strategy for discouraging consumption, such taxes will be regressive because poor households will bear a proportionately greater tax burden. Such issues are addressed in section 7.4 below.

Another approach to analyzing the effects of food prices and budget constraints on food consumption was proposed by Stigler (1945) using linear programming analysis. Calculation of the minimum costs of diets is helpful for those devising policies for meeting individuals' energy and nutrient requirements. Recently, the linear programming approach has become popular in the nutritional sciences (e.g. Darmon *et al.*, 2002). In addition to cost constraints, the objective function is minimized subject to an overall energy constraint, as well as "palatability" constraints, which include constraints on food quantities and the energy derived from each food group. For example, Darmon *et al.*, (2002) found that the higher prices of nutritious foods in France lowered the quality of diets, as reflected in reduced consumption of fruits and vegetables. However, food prices used in the analysis were averages for French regions from different time points. Using actual data on food prices paid by NFSPS households, Bhargava and Amialchuk (2007) found that use of "added sugars" increased (i.e. diet quality worsened) with more stringent cost constraints. However, linear programming analyses are mainly suggestive of households' food purchasing behavior, though by imposing additional constraints, one can embody the salient features of food consumption among populations.

Furthermore, the income elasticities of food groups and nutrients can provide insights into the effects of incomes on food consumption patterns. For example, Yen *et al.* (2003) found that in the NFSPS data, the use of fruits and vegetables increased with incomes. Moreover, for the NFSPS data, Bhargava and Amialchuk (2007) reported significant positive effects of households' incomes on the consumption of iron, folic acid, and vitamins B6 and B12. However, the interpretation of income elasticities of nutrients was limited by the study design, since all households in the NFSPS had incomes below certain thresholds of qualification for food stamp benefits. Even so, as seen in section 2.3 for developing countries, diet quality is likely to be worse for low-income households. The interesting contrast is that poor households in developed countries can meet their energy needs, though the food groups selected are likely to reduce the intake of nutrients, such as vitamins A and C, that are critical for maintaining health. Moreover, in developed countries, dietary knowledge and behavioral factors play an important role in selecting healthy foods from the vast number of foods available. Such aspects have not been incorporated in linear programming analyses, which have focused primarily on the effects of prices and household incomes on food consumption.

7.3.2 Some studies linking economic variables to the prevalence of obesity in developed countries

Economists have emphasized variables such as low food prices and the increased opportunity cost of time spent on cooking as factors contributing to the obesity epidemic. In a survey, Cutler *et al.* (2003) investigated data sets for the US and other developed countries in order to explain increases in obesity during the past few decades. One of their main arguments is that food processing techniques are now centralized so that food can be produced in large quantities at certain locations and distributed efficiently to other parts of the country. Moreover, partly because of easy access to processed and inexpensive foods, individuals consume snacks more frequently. Thus, individuals' time allocation patterns and dietary habits would not seem conducive to maintaining energy balance; energy expenditures are likely to decrease, while energy intakes have increased due to over-consumption of processed foods. Cutler *et al.* emphasized increases in food intakes and, on the basis of time-use data from Robinson and Godbey (1997), argued that energy expenditures have not markedly declined in the past three decades. By contrast, Prentice and Jebb (1995) have argued that energy expenditures in the UK have declined in the last few decades, though food intakes have been relatively stable. While it is not possible to measure the energy intakes and expenditures of a large number of individuals over extended periods, increases in obesity in countries such as the US and UK are no doubt indicative of positive energy imbalances.

Cutler *et al.* (2003), using data from the Continuing Survey of Food Intakes (USDA, 2003) in 1977–8 and 1994–6, argued that energy intakes at meals such as lunch and dinner have been relatively stable over time. Moreover, Cutler *et al.* asserted that portion sizes in restaurants cannot have increased in recent years because of this stability of energy intakes by meal. However, it is often difficult to assess the energy and nutrient content of meals served in restaurants; making food more palatable and appealing to customers can entail a greater use of fats for cooking and garnishing. Cutler *et al.* offered the example of potato consumptions which has seen marked changes from the 1960s because French fries are now the most common form of potato intake. Even in the case of French fries, however, frying in oil increases energy content, so that food preparation in restaurants is likely to be a contributor to increases in energy intakes.

The emphasis of Cutler *et al.* (2003) on innovations in food preparation technologies and food availability is well-placed, though increases in obesity have multi-dimensional causes. For example, the greater frequency of dining in restaurants and consuming pre-cooked meals at home, low levels of physical activity and snacking while watching television are likely to play an important role in promoting obesity. However, from a policy standpoint, it is difficult to devise economic policies that can reduce energy intakes and increase energy expenditures. For example, one cannot tax pre-prepared foods with high fat content to decrease their consumption; many consumers with good dietary knowledge can select lower quantities of such foods and augment the meal with fresh fruits and vegetables. Moreover, individuals from poor households often hold more than one job, and lack of time for cooking can force them into consuming processed foods. Thus, taxing pre-prepared foods and snacks would disproportionately affect the poor. In contrast, there is a stronger case for subsidizing fresh fruits and vegetables by encouraging grocery stores to discount fresh (or one-day-old) fruits and vegetables for the poor receiving food aid (Bhargava and Amilachuk, 2007).

Last, it is worth mentioning two further economic studies, one linking maternal employment to childhood obesity in the US (Anderson *et al.*, 2003), and the other linking food prices in restaurants to increased obesity (Chou *et al.*, 2004). Anderson *et al.* (2003) used data from the National Longitudinal Survey of Youth (NLS-Y) and found that intensity of maternal employment, reflected in hours worked per week, was a positive and significant predictor of children's chances of being overweight and obese. This finding may have been due to the lower energy density of breast milk and also the better care provided by mothers taking time off to be with young children. However, Anderson *et al.* mainly explained children's chances of being overweight and obese; body weight (controlling for height) or BMI might be more suitable continuous dependent variables, especially for addressing problems of endogeneity. Moreover, children-specific unobserved characteristics were only addressed in models using changes in maternal employment hours. In such models, maternal employment was often not a significant predictor of the chances of being overweight or obese (Anderson *et al.*, 2003: table 3). Because long working hours are not conducive to breast-feeding, the implications of these results are that mother-friendly practices in the US, such as longer maternity leave, would be helpful in reducing childhood obesity. This is the case in European countries, especially in Scandinavia, where maternity benefits can extend to three years. Moreover, the authors could have been more cautious in suggesting that their findings had a "causal" interpretation; children's weights in the NLS-Y were mostly reported by mothers, and important confounding factors such as child care practices were approximate measures.

In a study analyzing data from the Behavioral Risk Factor Surveillance System (BRFSS) during the period 1984–99, Chou *et al.* (2004) explained individuals' BMI and chances of being obese by explanatory variables such as prices in fast food and full-service restaurants, the price of food consumed at home, cigarette prices, and background variables. The authors found that prices in restaurants had quadratic effects on BMI and on chances of obesity, though the authors qualified this finding by noting that the evidence suggested that the growth in restaurants was partly due to increases in the value of households' time. However, the authors pooled the data from cross-sectional surveys for various years, making their sample sizes very large (greater than one million). In such circumstances, standard errors were likely to be small, thereby increasing the chances that the coefficients had statistical significance. Other methods, such as creating "pseudo-panels" based on repeated cross-sectional data, might have been more appropriate. Furthermore, Chou *et al.* (2004) attributed substantial weight increases to reductions in the prevalence of smoking. Because their data were not longitudinal, the estimated magnitudes of the coefficients of smoking prevalence need to be interpreted with caution. In fact, in a study using longitudinal data from the Framingham Offspring Study, Bhargava (2003b) found that the elasticity of body weight with respect to the number of cigarettes smoked was 0.01, which is quite small. Thus, quitting smoking can lead to some gain in body weight, but not on such as large scale as persistent over-consumption of energy-dense foods in restaurants and at home.

7.4 Conclusion: a broader approach to stemming the obesity epidemic

The causes underlying the obesity epidemic in industrialized countries are multi-dimensional and complex. Economic factors such as low food prices, the affordability of pre-cooked meals at home, and inexpensive food in restaurants have increased overall dietary energy intakes. In addition, cultural and socio-economic factors influence the types of foods consumed during family meals. Psychological factors, such as individuals' dietary knowledge, self-efficacy and the perceived health effects of foods are important in making dietary choices. Moreover, support from family members when adopting healthy lifestyles is important for the long-term success of nutrition education programs. Last, energy and nutrient intakes are important for explaining the balance of energy, which is adversely affected by a decline in physical activity. The consumption of energy-dense foods has increased and energy absorption rates from such foods are high, due to improvements in food storage technology and safety regulations. In the absence of a concerted effort from nutritional and social scientists, therefore, it will be difficult to stem the obesity epidemic.

There are several implications of the discussion of the obesity epidemic for the formulation of food and health policies. First, evidence from the psychological and nutritional literatures has underscored the importance of increasing individuals' nutritional knowledge and improving dietary behaviors via nutrition education programs. While food consumption patterns have evolved gradually over time, the recent abundance of inexpensive, energy-dense foods needs to be confronted via education and self-control. This is especially important for low-education and low-income households in developed countries which have experienced food shortages in the past. In the US, for example, studies have shown that Hispanic mothers may not correctly perceive excessive body weight as a problem among children (Crawford *et al.*, 2004). Moreover, food labeling is an important instrument for disseminating nutritional information. While processed foods are often labeled for their energy, fat, sugar and sodium content, greater space should be devoted to outlining the adverse health effects of over-consumption.

Second, it would be helpful if public health agencies such as the US Department of Agriculture and the Department of Health and Human Services would develop a simplified food labeling system based on health effects. For example, the "Pick the Tick" program promoted by the New Zealand Heart Association has reduced salt consumption (Young and Swinburn, 2002). A numerical ranking system classifying various foods into health categories should be devised for most countries. For example, rank 1 would indicate the most desirable foods, while (say) rank 10 could denote the least essential foods. Fresh fruits and vegetables are nutrient-dense and should be given rank 1. Canned fruits should receive a lower rank depending on their added sugar content. At the other extreme, snacks high in salt and sugars should receive rank 10 since they increase energy intakes and contribute to chronic conditions (see also Popkin *et al.*, 2006). The ranking system should be widely publicized in government offices dealing with food stamps, and on the Internet, to ensure that consumers understand the rankings. Such a system would seem easier to comprehend than the US Department of Agriculture pyramid, which lumps food groups together and there is the frequent ambiguity in understanding the desirability of food groups.

Third, it is not sufficiently emphasized in the biomedical and public health sciences that individuals' energy requirements are strongly influenced by their body weights and by physical activity levels. Thus, the energy requirements of overweight and obese individuals are higher and it is easy for them to get trapped into high-energy-intake disequilibrium. By contrast, slim individuals are less likely to indulge in binge eating, since their systems are attuned to tighter bounds for energy intakes. Thus, it is imperative that physicians counsel individuals about weight loss at the earliest opportunity. This may not be feasible in countries such as the US, because a large number of uninsured individuals seek medical help only when complications associated with obesity, such as hypertension, diabetes and cardiovascular disease, are evident. Once these diseases develop, the lifelong costs of treatments are high; medical counseling at earlier stages is therefore likely to be cost-effective. Moreover, individuals' economic productivity is likely to be enhanced by prevention strategies. While actual gains in productivity may be small, better health improves the quality of life. Physical activity is also essential for avoiding weight gain and for improving profiles of the "good" HDL cholesterol (Bhargava, 2003b). Increases in exercise can be achieved by subsidizing facilities such as gyms, especially in poor neighborhoods.

Fourth, economic incentives and policies can help reduce excessive body weight, though it is important to take the welfare aspects into account. As noted in section 7.3.2, taxing non-nutritious foods may not be practicable, though there is a good case for subsidizing the consumption of fresh fruits and vegetables. Grocery stores can discount fresh (or day-old) fruits and vegetables for those receiving food aid, and such policies will ensure a greater supply of micronutrients for children in such households. Moreover, agricultural subsidies received by large-scale farms in developed countries keep prices of unhealthy foods, such as corn syrup, artificially low. Such policies not only promote the over-consumption of non-essential foods but also destroy the livelihoods of farmers in developing countries, who cannot compete with large-scale subsidized farming (Pinstrup-Andersen, 2002; Elinder, 2005). While farmers growing fresh fruits and vegetables in developed countries may occasionally require subsidies to stabilize prices, it is illogical to lower the costs of agribusinesses promoting non-essential foods to individuals who are unaware of the adverse health effects. Thus, economic analyses of agricultural markets should focus on the production of foods from the standpoint of their desirability for health outcomes, and investigate circumstances under which subsidies are necessary for the production of foods such as fruits and vegetables.

Finally, "all-you-can-eat"-type restaurants can promote overeating; such restaurants are increasing in countries such as the US, presumably due to the higher energy requirements of overweight individuals. Profit margins in these restaurants are likely to be small and hence the owners need to attract greater numbers of customers. Thus, general guidelines should be developed for restaurants in order to discourage overconsumption of energy-dense food. For example, more detailed information should be provided to customers on the fat and sugar content of foods served. Moreover, the energy content of meals bundling foods such as hamburgers with fries should be reported on the menus. However, due to the large numbers of overweight individuals with low education levels in the US, it is likely that only a small proportion would benefit from such regulations. Thus, to stem the obesity epidemic, it would seem essential to increase nutrition education and physical activity programs, print detailed labels on foods sold in grocery stores, and provide subsidies for individuals participating in weight-loss programs. Furthermore, it would be useful for governments to tax advertising of non-nutritious foods at high rates and use the proceeds for nutrition education programs, which are expensive to implement. Without a concerted effort from public health and social scientists and governmental agencies, the health of the population and its quality of life are likely to decline because of the obesity epidemic.

8

Summing up and concluding remarks

This monograph covered several concepts from disciplines such as economics, nutrition, psychology, demography, epidemiology and public health, and presented empirical evidence for the formulation of food and health policies. While the issues are undoubtedly complex, policy design in developing and developed countries demands multi-disciplinary approaches. It is hoped that readers will find the material to be a useful summary of important issues that are analyzed within and outside their respective fields. This chapter reiterates the major themes and findings from the book and points to the need for further research on food policy issues.

The demand for food and nutrients in developing countries was studied in Chapter 2, since it is important to investigate how dietary intakes change with household incomes. For example, if diet quality, reflected in the intakes of micronutrients such as calcium, iron, and vitamins A and C, improves with incomes, then policy-makers need not be too concerned with the design of food policies, especially in countries experiencing economic growth. Because food expenditures data from household surveys are in an aggregate form, the income elasticities of energy and nutrients estimated from such data sets in the previous economics literature were likely to be higher than the corresponding estimates from food intakes data compiled using 24-hour recall or other direct methods. Thus, households' intakes of energy and nutrients are likely to increase at a slower rate than that predicted by income elasticities from food expenditures data. Moreover, as emphasized in the nutrition and physiology literatures (FAO/UNU/WHO, 1985), it is important to account for individuals' energy expenditures in empirical models. Models estimated in the economics literature focus mainly on the role of household incomes and food prices. The econometric models presented in Chapter 2 incorporated the physiological and economic aspects and should therefore be useful for food policy analysis.

The empirical results in Chapter 2, based on food intakes data from India, the Philippines, Kenya and Bangladesh, showed income elasticities of energy to be in the neighborhood of 0.10; the income elasticity of energy intakes was rather high for Kenya (0.39), where during the survey period there were food shortages due to a drought. The income elasticities of micronutrients in these countries ranged from 0.02 to 0.42 and were high in the Philippines and Kenya. While high estimates of income elasticities are desirable because economic development will improve nutrient intakes, widespread poverty may suggest the need for interventions in the form of direct or indirect food subsidies to households. This is especially important for vitamin A and C intakes, which are affected by seasonal variations in fruit and vegetable prices, and for calcium and iron intakes, because absorption of these nutrients is affected by other nutrients present in the meal. Adequate protein intakes are important for maintaining the growth of children and for the labor productivity of adults.

From a food policy perspective, energy deficiencies in developing countries have declined in the past few decades due to improvements in production technologies; however, protein and micronutrient deficiencies merit greater attention. By incorporating the knowledge from biomedical sciences into econometric models for energy and nutrient intakes, researchers can formulate policies for improving the nutritional status of inhabitants of developing countries. The income elasticities of nutrients presented in Chapter 2 provide guidance for designing interventions, especially when nutrient intakes respond slowly to increases in incomes. For example, policy-makers need to ensure higher intakes of dairy products and fruits and vegetables. Such interventions include schemes for increasing milk and meat production, and distributing improved varieties of seeds for enhancing the vitamin and mineral content of staple foods and vegetables. The benefits of such interventions are likely to reach large numbers of individuals and, in the long run, will improve aggregate indicators of population health such as life expectancy. Moreover, the costs associated with agricultural technologies continue to fall and policy-makers need to devise appropriate strategies for countries, depending on climatic conditions and irrigation methods. Further research on nutrient deficiencies among population subgroups and their effects on health outcomes can guide the channeling of resources for improving diet quality in developing countries.

In Chapter 3, the effects of socio-economic variables and nutritional intakes on indicators of child health such as height, weight and sicknesses were investigated using data from developing countries. The analytical framework was quite general and, as shown later in Chapter 7, can be extended for analysing the proximate determinants of obesity. While nutritional intakes in developing countries improved with rises in household incomes (Chapter 2), it is important to investigate the effects of specific nutrients on indicators of child health. For example, higher intakes of vitamins A and C predicted lower child morbidity in the Philippines and Kenya. This evidence strengthens the case in Chapter 2 for policies that make fruits and vegetables more affordable for poor households in developing countries.

From a methodological standpoint, the framework for investigating the role of socio-economic, nutritional and environmental factors on indicators of child health combined the approaches in the biomedical and social sciences. The emphasis in the economics literature on the effects of food prices on health indicators is perhaps somewhat narrow, especially when most households in the surveys reside in the same geographical area and face similar prices. However, econometric techniques are useful for tackling the endogeneity of explanatory variables in models for health relationships. Moreover, by invoking assumptions that are consistent with the biomedical sciences, econometric approaches can encompass the anthropometric assessment literature on the relationships between variables such as heights and weights. While food prices affect food consumption, the effects of food prices on health indicators are likely to operate via specific dietary intakes. For example, policy-makers are interested in knowing the effects of intakes of protein, calcium and iron on child growth, and econometric models and empirical evidence presented in Chapter 3 should be useful for formulating food policies.

The importance of diet quality was evident in the models for Filipino children, where protein and calcium intakes were significant predictors of heights; β-carotene intakes were associated with lower morbidity. Moreover, the protein–energy ratio was a significant predictor of the weights of Kenyan children. The empirical models also highlighted the role for children's morbidity of environmental factors such as sanitation; the presence of coliforms in the water in Bangladesh increased child morbidity. It is important that food policies take a broad view of health in developing countries and enable children to achieve their full physical and cognitive potential. It is also useful jointly to assess the costs of various interventions, and the formulation of policies would benefit from long-term considerations. For example, poor sanitation increases the transmission of bacterial diseases and entails nutrient loss via sicknesses. While it is expensive to replenish the human body's nutrient stores, sewage treatment is an expensive intervention that is feasible mainly in middle- and high-income countries. However, policy-makers need to take a long-term view of the health benefits of cleaning up the environment. For example, developing countries aspiring to be economic powers, such as China and India, cannot afford to neglect waste disposal and sewage treatment. With increased prosperity, piecemeal approaches are affordable interventions. For example, better waste disposal methods can be devised relatively cheaply in the short run, and trust funds can be set up to cover the expected costs of sewage treatment in the future. The evidence in Chapter 3 pointed towards the need for greater investments in sanitation and hygiene to improve population health in developing countries.

Chapter 4 extended the framework in Chapter 3 for the analysis of health indicators to children's psychological indicators, such as scores on cognitive and educational achievement tests. Children's learning is a cumulative process and the future supply of skilled labor depends on the level of children's cognitive development. Developing countries where most children are without primary school education are unlikely to be competitive in the global economy because the goods and services exported by a competitive economy typically reflect high added value. While one can enhance the productivity of unskilled labor via improvements in the quality of their food intakes (Chapter 6), it takes much longer to educate children and train them for employment in skilled occupations. For the formulation of food and health policies, it is important to understand the mechanisms underlying children's cognitive development and to incorporate the empirical evidence from developing countries.

The analyses of data on Kenyan infants showed the importance of maternal nutritional and health status for infants' length, weight and head circumference at birth and for growth in these indicators in the first six months. Diet quality, reflected in protein intakes from animal sources, was a predictor of the scores on the Bayley Infant Behaviour Record at six months. Furthermore, results from the analyses of data on Kenyan and Tanzanian schoolchildren showed the importance of children's health status for their scores on cognitive and educational achievement tests. The empirical results identified key determinants of child development, such as height, weight, and hemoglobin concentration, that were predictors of scores on cognitive and educational achievement tests. It is evident that maternal nutritional status in developing countries needs be improved via food policies for promoting child growth. For example, iron supplementation for pregnant and lactating women and supplemental food high in animal products in areas of food shortages are critical strategies for ensuring that young children avoid growth retardation. Because physical growth and cognitive development are intertwined for children growing up in poverty, food policies that improve diets by increasing intakes of protein, iron, calcium, and vitamins A and C are likely to have beneficial long-term effects on children's learning and skill acquisition. The availability of skilled labor, in turn, is important for economic growth in developing countries (Bhargava, Jamison *et al.*, 2001).

In addition, households' socio-economic status was an important predictor of school attendance among Tanzanian children. Thus, subsidies for school fees and supplies, especially in the wake of the AIDS epidemic in sub-Saharan Africa (Bhargava and Bigombe, 2003), are likely to increase the school participation of children from poor households. From a methodological standpoint, analyses for Tanzanian schoolchildren were based on the data from a randomized controlled trial. While such data sets enable useful comparisons that may not be feasible using observational data, it was pointed out that interventions in social science settings can have unforeseen effects on the variables of interest. For example, parents may ensure that children attend school on days of interventions so as to receive health benefits from participating in the control or intervention group. Social scientists need to be careful in assessing the benefits of interventions for health and other outcomes. In particular, the data from randomized controlled trials should be analyzed in accordance with the likely pathways linking variables changed by the intervention to specific outcomes. For example, treatment for intestinal parasites such as hookworm is likely to increase children's hemoglobin concentration. But there may not be systematic relationships between deworming and school attendance unless de-worming reduces morbidity spells and improves overall health status, which, in turn, enables children to attend school regularly. Multi-disciplinary analyses are useful for the formulation of food, education and health policies since they take into account diverse factors affecting child development. In contrast, focusing on a limited set of factors can lead to suboptimal resource allocation, since important variables analyzed in other disciplines may be missing from the empirical models. Overall, the formulation of efficacious food policies for developing countries requires the incorporation of evidence from various sources, including randomized controlled trials, household surveys and macro-economic analyses.

Chapter 5 addressed issues of high fertility (number of children born) and child mortality, which are a focus of demographic research in developing countries. These aspects are critical for economic development because the health and education of surviving children critically depend on family resources, which decline with additional births. Moreover, high fertility rates in rural areas of developing countries and limited work opportunities increase the migration of unskilled labor to urban areas, leading to congestion and the development of slums. Demographic aspects are not sufficiently emphasized in food policy research, even though policies promoting sustainable economic development critically depend on low fertility rates and the available natural and human resources. Even in developing countries with ample natural resources such as those with oil reserves in the Middle East, lower fertility rates can facilitate children's education and promote political stability. Both these factors are key elements of sustainable economic development.

Several aspects of the relationships between fertility and child mortality in developing countries were discussed in Chapter 5, starting with possible discrimination against girls. New evidence on the heights and weights of boys and girls from Pakistan and Vietnam was presented; using the NFHS-1 data from India, it was argued that child mortality was a more reliable indicator for investigating the selective neglect of girls. Furthermore, the role of the health care infrastructure in lowering fertility and child mortality rates was underscored. While social scientists have emphasized female education as a means of lowering fertility rates, there are time lags of 6 to 10 years between educating girls and their reaching reproductive ages. Moreover, the education of girls can be hampered by large numbers of siblings, especially in poor households. Thus, it is essential that governments in developing countries invest in the health care infrastructure to increase the uptake of family planning services. This will enable mothers to devote more time to child care and will be beneficial for the health of surviving children and their cognitive development (Chapter 4). Moreover, countries such as Japan and South Korea, which are not endowed with many natural resources, have prospered by lowering fertility rates and educating children. Thus, future research should pay greater attention to the effects of high fertility rates on economic development, especially in countries in sub-Saharan Africa and Asia, where large numbers of individuals are trapped in poverty due to lack of education and economic opportunities.

Chapter 6 discussed the effects of the nutritional and health status of adults on labor productivity in developing countries, which is important for improving the living standards of wage earners and their families. Under-nourished adults may not be able to perform strenuous tasks that are required in unskilled occupations, thereby increasing the chances of households falling into poverty. It was seen in Chapter 2 that increases in household incomes led to improvements in diet quality in countries such as India, the Philippines and Kenya. In contrast, Chapter 6 was concerned with income gains from better nutritional and health status. The evidence from biomedical studies indicates beneficial effects of nutritional status on physical work capacity and productivity. However, demonstrating these effects in population settings is complex. While wages earned are expected to reflect productivity, wages are influenced by macroeconomic activity, such as labor market conditions. Moreover, a majority of rural households in developing countries do not receive payments in cash for all their services. Thus, it is important to analyze the proximate determinants of time allocated to productive and subsistence activities in developing countries.

The economics literature also emphasizes the effects of nutrition on wages and productivity. However, it was argued in Chapter 6 that nutrient intakes gradually improve individuals' health status and that employers are likely to offer higher wages to workers that appear to be well nourished. Thus, the onus of improving the health status of poor adults may fall on national governments and international agencies, and food policies are crucial for improving labor productivity. Because of uncertainty in the factors underlying wages earned, the analyses of time allocation data from Rwanda provided many useful insights. For example, it was seen that adults with high BMI spent a smaller proportion of time resting, sleeping and sitting quietly, and a greater proportion of time on heavy activities demanding high energy expenditures. Moreover, average energy expenditures were around twice the BMR and the adults lost weight during the survey period. Thus, food supplementation programs are likely to enhance health status and labor productivity in countries such as Rwanda. In addition, micronutrient deficiencies can be identified from nutritional surveys, and food policies that alleviate nutrient deficiencies can raise labor productivity. For example, increasing the intakes of protein and iron from animal sources is likely to enhance performance on agricultural and heavy activities. Improvements in health status and life expectancy are important predictors of GDP growth rates in low-income countries (Bhargava, Jamison *et al.*, 2001). Such investments are likely to

185

be self-sustaining, especially if fertility rates are reduced via health care interventions to facilitate children's education (Chapter 5).

Chapter 7 was concerned with issues of obesity in developed countries such as the US. The obesity epidemic is also affecting affluent sections of societies in middle- and low-income countries, since food expenditures are a small proportion of those households' budgets. Moreover, food prices are low partly because of agricultural subsidies to farmers in developed countries. Individuals with poor dietary knowledge and self-control are likely to over-consume food, and maintaining a healthy body weight is difficult among affluent populations. Moreover, the costs of treatment of chronic conditions associated with obesity are high and so policy-makers urgently need to devise policies for dealing with the obesity epidemic.

There are several implications of the discussion in Chapter 7 for food and health policies. First, it is important to increase individuals' nutritional knowledge and improve diets via nutritional education programs. These are costly interventions in developed countries but are essential for stemming the obesity epidemic. Some of the funds necessary for such programs may be raised by taxing advertising by manufacturing companies of non-nutritious foods. Second, public health agencies such as the US Department of Agriculture and the Department of Health and Human Services should develop simplified food labeling systems based on health effects. For example, as outlined in Chapter 7, rank 1 would indicate the most desirable foods, while rank 10 could denote the least essential foods. Fresh fruits and vegetables should be given top rank 1. The ranking system should be widely publicized in offices administering food supplementation programs. Third, it is not emphasized enough in the biomedical and public health sciences that individuals' energy requirements are strongly influenced by their body weights and physical activity levels. Thus, the energy requirements of obese individuals are higher and it is imperative that physicians counsel such individuals about weight loss at the earliest opportunity. Physical activity is essential for weight loss, which in turn decreases energy requirements and the need to consume energy-dense foods, which are often unhealthy.

Fourth, economic incentives and policies can help reduce the prevalence of obesity, though it is important to take welfare aspects into account. As noted in Chapter 7, taxing non-nutritious foods may not be practicable, though there is a case for subsidizing the consumption of fresh fruits and vegetables. Moreover, agricultural subsidies received by large-scale farms in developed countries keep prices of unhealthy foods such as corn syrup artificially low. Such policies not only promote over-consumption of non-essential foods but also destroy the livelihoods of farmers in developing countries (Pinstrup-Andersen, 2002; Elinder, 2005). Last, "all-you-can-eat"-type restaurants can promote overeating and guidelines should be developed for restaurants to discourage over-consumption of energy-dense food. It would seem essential to expand nutrition education and physical activity programs for low-income populations in developed countries; detailed labels on foods sold in grocery stores and incentives for participating in weight-loss programs are important strategies for combating the obesity epidemic.

In concluding this monograph, it might be added that with respect to the design of food and health policies, the demarcations between disciplines such as economics, nutrition, psychology, epidemiology and public health are superficial and may even be counter-productive. Due to the technical nature of the issues, however, researchers often focus on aspects emphasized in their disciplines. Even so, recognition of research in other disciplines is helpful for policy-makers utilizing the empirical results. For example, there is usually agreement between the approaches in the nutritional sciences and epidemiology on how to tackle the obesity epidemic; social scientists can forge similar links to broaden the scope of analyses, since food consumption depends on economic and cultural factors. However, a key difference between the biological and social sciences approaches is that social scientists invoke assumptions that are often untested and are often context-specific. In contrast, assumptions in the biological sciences are based on evidence from experimental studies and can be invoked under similar circumstances in other situations. Because of the recent popularity of randomized controlled trials in the social sciences, multi-disciplinary research can narrow the differences between disciplines, provided that researchers agree on a set of assumptions that reflect salient aspects of the problems under study. While additional methodological issues arise in analyses of the data from randomized controlled trials, multi-disciplinary teams are an efficient way of conducting and disseminating research for the design of food and health policies.

References

Adams, P., Hurd, M., McFadden, D., Merrill, A., and Ribeiro, T. (2003), "Healthy, wealthy, and wise? Tests for direct causal paths between health and socioeconomic status", *Journal of Econometrics*, 112: 3–133.

Al, M., Houwelingen, A., and Hornstra, G. (1997), "Relation between birth order and the maternal and neonatal docohexaenoic acid status", *European Journal of Clinical Nutrition*, 51: 548–53.

Alderman, H. (1986), *Cooperative Dairy Development in Karnataka, India: An Assessment*, Research Report 64 (Washington, D.C.: International Food Policy Research Institute).

Anderson, P., Butcher, K., and Levine, P. (2003), "Maternal employment and overweight children", *Journal of Health Economics*, 22: 477–504.

Angeles, G., Guilkey, D., and Mroz, T. (1998), "Purposive program placement and the estimation of family planning program effects in Tanzania", *Journal of the American Statistical Association*, 93: 884–99.

Ashenfelter, O., and Heckman, J. (1974), "The estimation of income and substitution effects in a model of family labor supply", *Econometrica*, 42: 73–85.

Astrand, P., and Rodahl, K. (1986), *Textbook of work physiology* (New York: McGraw Hill).

Aturupane, H., Glewwe, P., and Isenman, P. (1994), "Poverty, human development and growth: An emerging consensus", *American Economic Review*, 84: 344–9.

Bandura, A. (1977), *Social Learning Theory* (Englewood Cliffs, N.J.: Prentice Hall)

—— (1986), *Social Foundations of Thought and Action* (Englewood Cliffs, N.J.: Prentice Hall).

Baranowski, T., Cullen, K., Nicklas, T., Thompson, D., and Baranowski, J. (2002), "School-based obesity prevention: A blueprint for taming the epidemic", *American Journal of Health Behavior*, 26: 486–93.

Basta, S. S, Soekirman, M. S., Karyadi, D., and Scrimshaw, N. S. (1979), "Iron deficiency anemia and the productivity of adult males in Indonesia", *American Journal of Clinical Nutrition*, 32: 916–25.

Bayley, N. (1969), *Bayley Scales of Infant Development* (New York: Psychological Corporation).

Beard, J. (1995), "Effectiveness and strategies of iron supplementation during pregnancy", *American Journal of Clinical Nutrition*, 71: 1288S–1294S.

Beaton, G. H. (1984), "Adaptation to and accommodation of long term low energy intake", in E. Pollitt and P. Amante (eds), *Energy Intake and Activity* (New York: Alan R. Liss, Inc.), 395–403.

Becker, G. (1965), "A theory of the allocation of time", *Economics Journal*, 75: 493–517.

Becker, G. (1982), "A theory of competition among pressure groups for political influence", *Quarterly Journal of Economics*, 98: 371–400.

Becker, G. S. (1991), *A Treatise on the Family* (Cambridge, Mass.: Harvard University Press).

Behrman, J., and Deolalikar, A. (1987), "Will developing country nutrition improve with incomes? A case study for rural south India", *Journal of Political Economy*, 95: 492–507.

—— —— (1988), "Health and nutrition", in H. Chenery and T. N. Srinivasan (eds), *Handbook of Development Economics* (Amsterdam: Elsevier Science), 631–771.

Berio, A.-J. (1984), "The analysis of time allocation and activity patterns in nutrition and rural development planning", *Food and Nutrition Bulletin*, 6: 53–68.

Bhargava, A. (1991a), "Estimating short and long run income elasticities of foods and nutrients for rural south India", *Journal of the Royal Statistical Society A*, 154: 157–74.

—— (1991b), "Identification and panel data models with endogenous regressors", *Review of Economic Studies*, 58: 129–40.

—— (1992), "Malnutrition and the role of individual variation with evidence from India and the Philippines", *Journal of the Royal Statistical Society A*, 155: 221–31.

—— (1994), "Modelling the health of Filipino children", *Journal of the Royal Statistical Society A*, 157: 417–32.

—— (1997), "Nutritional status and the allocation of time in Rwandese households", *Journal of Econometrics*, 77: 277–95.

—— (1998), "A dynamic model for the cognitive development of Kenyan schoolchildren", *Journal of Educational Psychology*, 90: 162–7.

—— (1999), "Modelling the effects of nutritional and socioeconomic factors on the physical development and morbidity of Kenyan school children", *American Journal of Human Biology*, 11: 317–26.

—— (2000), "Modeling the effects of maternal nutritional status and socioeconomic variables on the anthropometric and psychological indicators of Kenyan infants from age 0–6 months", *American Journal of Physical Anthropology*, 111: 89–104.

—— (2001a), "Nutrition, health and economic development: Some policy priorities", *Food and Nutrition Bulletin*, 22: 173–7.

—— (2001b), "World Health Report 2000", *Lancet*, 358: 1097–8.

—— (2003a), "Family planning, gender differences and infant mortality: Evidence from Uttar Pradesh, India", *Journal of Econometrics*, 112: 225–40.

—— (2003b), "A longitudinal analysis of the risk factors for diabetes and coronary heart disease in the Framingham Offspring study", *Population Health Metrics*, 14 Apr.: 1–10.

References

Bhargava, A. (2004), "Socio-economic and behavioural factors are predictors of food use in the National Food Stamp Program Survey", *British Journal of Nutrition*, 92: 497–506.

—— (2005), "AIDS epidemic and the psychological well-being and school participation of Ethiopian orphans", *Psychology, Health and Medicine*, 10: 263–76.

—— (2006a), "Econometrics, statistics and computational approaches in food and health sciences" (Singapore: World Scientific Publishing Company), available at <www.worldscibooks.com/economics/6140.html>.

—— (2006b), "Fiber intakes and anthropometric measures are predictors of circulating hormones, triglyceride and cholesterol concentrations in the Women's Health Trial", *Journal of Nutrition*, 136: 2249–56.

—— (2007), "Desired family size, family planning and fertility in Ethiopia", *Journal of Biosocial Science*, 39: 367–81.

—— and Amialchuk, A. (2007), "Added sugars displaced the use of vital nutrients in the National Food Stamp program survey", *Journal of Nutrition*, 137: 453–60.

—— and Bigombe, B. (2003), "Public policies and the orphans of AIDS in Africa", *British Medical Journal*, 326 (7403): 1387–9.

—— and Fox-Kean, M. (2003), "The effects of maternal education versus cognitive test scores on child nutrition in Kenya", *Economics and Human Biology*, 1: 309–19.

—— and Guthrie, J. F. (2002), "Unhealthy eating habits, physical exercise and macronutrient intakes are predictors of anthropometric indicators in the Women's Health Trial: Feasibility Study in Minority Populations", *British Journal of Nutrition*, 88: 719–28.

—— and Hays, J. (2004), "Behavioral variables and education are predictors of dietary change in the Women's Health Trial: Feasibility study in minority populations", *Preventive Medicine*, 38: 442–51.

—— and M. Ravallion, (1993), "Does household consumption behave as a martingale? A test for rural South India", *Review of Economics and Statistics*, 76: 500–4.

—— and Reeds, P. J. (1995), "Requirements for what? Is the measurement of energy expenditure a sufficient estimate of energy needs?", *Journal of Nutrition*, 125: 1358–62.

—— and Sargan, J. D. (1983), "Estimating dynamic random effects models from panel data covering short time periods", *Econometrica*, 51: 1635–59.

—— and Yu, J. (1997), "A longitudinal analysis of infant and child mortality rates in developing countries", *Indian Economic Review*, 32: 141–53.

—— Franzini, L., and Narendranathan, W. (1982), "Serial correlation and the fixed effects model", *Review of Economic Studies*, 49: 533–49.

—— Forthofer, R., McPherson, S., and Nichaman, M. (1994), "Estimating the variations and autocorrelations in dietary intakes on weekdays and weekends", *Statistics in Medicine*, 13: 113–26.

—— Bouis, H. E., and Scrimshaw, N. S. (2001), "Dietary intakes and socioeconomic factors are associated with the hemoglobin concentration of Bangladeshi women", *Journal of Nutrition*, 131: 758–64.

—— Jamison, D. T., Lau, L. J., and Murray, C. J. (2001), "Modeling the effects of health on economic growth", *Journal of Health Economics*, 20: 423–40.

—— —— and Hallman, K., and Hoque, B. A. (2003), "Coliforms in the water and hemoglobin concentration are predictors of gastrointestinal morbidity of Bangladeshi children ages 1–10 years", *American Journal of Human Biology*, 15: 209–19.

—— Jukes, M., Lambo, J., Kihamia, C. M., Lorri, W., Nokes, C., *et al* (2003), "Anthelmintic treatment improves the hemoglobin and serum ferritin concentrations of Tanzanian schoolchildren", *Food and Nutrition Bulletin*, 24: 332–42.

—— —— Ngorosho, D., Khilma, C., and Bundy, D. (2005), "Modeling the effects of health status and the educational infrastructure on the cognitive development of Tanzania school children", *American Journal of Human Biology*, 17: 280–92.

—— Chowdhury, S., and Singh, K. K. (2005), "Healthcare infrastructure, contraceptive use and infant mortality in Uttar Pradesh, India", *Economics and Human Biology*, 3: 388–404.

Bhutta, Z., Darmstadt, G., Hasan, B., and Haws, R. (2005), "Community-based interventions for improving perinatal and neonatal health outcomes in developing countries: A review of the evidence", *Pediatrics*, 115: 519–617.

Binet, A., and Simon, T. (1916), *The Development of Intelligence in Children* (Baltimore: Williams and Wilkins).

Bingham, S. A. (1994), "The use of 24-hour urine samples and energy expenditures to validate dietary assessment", *American Journal of Clinical Nutrition*, 59: 227S–31S.

Binswanger, H., and Jodha, N. (1978), *Manual for Instruction for Economic Investigators in ICRISAT's Village Level Studies* (Hyderabad: International Crops Research Institute for Semi-arid Tropics).

Blackburn, G. L., Chlebowski, R. T., Elashoff, R. M., and Wynder, E. L. (1996), "Monitoring dietary change in a low-fat diet intervention study: Advantages of using 24-hour dietary recalls versus food records", *Journal of the American Dietetic Association*, 96: 574–9.

Bliss, C., and Stern, N. (1978), "Productivity, wages and nutrition, I and II", *Journal of Development Economics*, 5: 331–62 and 363–98.

Block, G. (1982), "A review of validation of dietary assessment methods", *American Journal of Epidemiology*, 115: 492–505.

Bongaarts, J. (1990), "The measurement of wanted fertility", *Population Development Review*, 16: 487–506.

—— (1993), "The supply–demand framework for the determinants of fertility: An alternative implementation", *Population Studies*, 47: 437–56.

Bouis, H. E., and Haddad, L. (1992), "Are estimates of calorie-income elasticities too high?", *Journal of Development Economics*, 39: 333–64.

—— Briere, B., Guitierrez, L., *et al.* (1998), *Commercial vegetable and polyculture fish production in Bangladesh: Their impacts on income, household resource allocation, and nutrition"*, Washington, D.C.: International Food Policy Research Institute

References

Bowen, D., Clifford, C.C., Coates, R., *et al.* (1996), "The Women's Health Trial Feasibility Study in Minority Populations: Design and baseline description", *Annals of Epidemiology*, 6: 507–19.

Bray, G., and Popkin, B. (1998), "Dietary fat intake does affect obesity!", *American Journal of Clinical Nutrition*, 68: 1157–73.

Brazelton, T. (1984), *Neonatal Behavioral Assessment Scale*, 2nd edn (London: Blackwell Publishers).

Buzzard, I. M., Faucett, C. L., Jeffery, R. W., *et al.* (1996), "Monitoring dietary change in a low-fat diet intervention study: Advantages of using 24-hour dietary recalls vs food records", *Journal of the American Dietary Association*, 96: 574–9.

Caldwell, J. C. (1982), *Theory of Fertility Decline* (London: Academic Press).

Calloway, D. H., Murphy, S. P., and Bunch, S. (1994), *User's Guide to the International Minilist Nutrient Data Base* (Berkeley, Calif.: Department of Nutritional Sciences, University of California).

Campbell, K., Waters, E., O'Meara, S., and Summerbell, C. (2001), "Interventions for preventing obesity in childhood: A systematic review", *Obesity Reviews*, 2: 1–9.

Casanueva, E., Viteri, F., Mares-Galindo, M., *et al.* (2005), "Weekly iron as a safe alternative to daily supplementation for nonanemic pregnant women", *Archives of Medical Research*, 37: 674–82.

Cebu Study Team (1992), "A child health production function estimated from longitudinal data", *Journal of Development Economics*, 38: 323–51.

Ceci, S. J. (1991), "How much does schooling influence intelligence and its cognitive components? A reassessment of the evidence", *Developmental Psychology*, 27: 703–22.

Ceesay, S., Prentice, A., Cole, T., *et al.* (1997), "Effects on birth weight and perinatal mortality of maternal dietary supplements in rural Gambia: 5 year randomized controlled trial", *British Medical Journal*, 315: 786–90.

Center for Disease Control (2005), "A SAS program for the CDC Growth Charts", available at <www.cdc.gov/nccdphp/dnpa/growthcharts/sas.htm>.

Chandra, R., Au, B., Woodford, G., and Hyam, P. (1977), "Iron status, immunicompetence and susceptibility to infection", in A. Jacob (ed.), *Ciba Foundation Symposium on Iron Metabolism* (Amsterdam: Elsevier), 249–68.

Checkley, W., Epstein, L., Gilman, R., Cabrera, L., and Black, R. (2003), "Effects of acute diarrhea on linear growth of Peruvian children", *American Journal of Epidemiology*, 157: 166–75.

Chen, L., Huq, E., and D'Souza, S. (1981), "Sex bias in the family allocation of food and health in rural Bangladesh", *Population and Development Review*, 7: 50–77.

Chenery, H. (1949), "Engineering production functions", *Quarterly Journal of Economics*, 63: 507–31.

Chou, S.-Y., Grossman, M., and Saffer, H. (2004), "An economic analysis of adult obesity: Results from the Behavioral Risk Surveillance System", *Journal of Health Economics*, 23: 565–87.

Cleland, J. (1996), "Population growth in the 21st century: Cause for crisis or celebration?", *Tropical Medicine and International Health*, 1: 15–26.

——— and Wilson, C. (1987), "Demand theories of fertility transition: An iconoclastic view", *Population Studies*, 41: 5–30.

Coates, R., Bowen, D. J., Kristal, A. R., *et al.* (1999), "The women's health trial feasibility study in minority populations: Changes in dietary intakes", *American Journal of Epidemiology*, 149: 1104–12.

Cohen, B., Ohls, J., Andrews, M., *et al.* (1999), *Food Stamp Participants Food Security and Nutrient Availability*, (Washington, D.C.: Technical Report Economic Research Service, United States Department of Agriculture).

Cohen, J. M., and Lewis, D. B. (1987), "Role of government in combating food shortages: Lessons from Kenya 1984–85", in M. Glantz (ed.), *Drought and Hunger in Africa* (Cambridge: Cambridge University Press), 269–96.

Cole, T. (1991), "Weight–stature indices to measure under-weight, over-weight and obesity", in J. Himes (ed.), *Anthropometric Assessment of Nutritional Status* (New York:Wiley-Liss), 83–111.

Contento, I. (1995), "The effectiveness of nutrition education and implications of nutrition education policy, programs, and research: A review of research", *Journal of Nutrition Education*, 27: 279–418.

Cox, D. R. (1958), *Planning of Experiments* (New York: John Wiley and Sons).

——— (1970), *Analysis of Binary Data* (London: Chapman and Hall).

——— (1992), "Causality: Some statistical aspects", *Journal of the Royal Statistical Society A*, 155: 291–301.

——— and Snell, E. (1968), "A general definition of residuals (with discussion)", *Journal of Royal Statistical Society B*, 30: 248–275.

Crawford, P., Gosliner, W., Anderson, C., *et al.* (2004), "Counseling Latina mothers of preschool children about weight issues: Suggestions for a new framework", *Journal of the American Dietetic Association*, 104: 387–94.

Cronbach, L. J. (1984) *Essentials of psychological testing*, 4th edn (New York: Harper and Row).

Cutler, D., Glaeser, E., and Shapiro, J. (2003), "Why Americans have become more obese", *Journal of Economic Perspectives*, 17: 93–118.

Darmon, N., Ferguson, E., and Briend, A. (2002), "A cost constraint alone has adverse effects on food selection and nutrient density: An analysis of human diets by linear programming", *Journal of Nutrition*, 132: 3764–71.

Das Gupta, M. (1987), "Selective discrimination against female children in rural Punjab", *Population and Development Review*, 13: 77–100.

Dasgupta, P. (1993), *An Inquiry into Well-Being and Destitution* (New York: Oxford University Press).

Deaton, A., and Muellbauer, J. (1980), *Economics and Consumer Behaviour* (Cambridge: Cambridge University Press).

Deolalikar, A. B. (1988), "Nutrition and labor productivity in agriculture: Estimates for rural south India", *Review of Economics and Statistics*, 67: 406–13.

References

Dickson, R., Awasthi, S., Williamson, P., Demellweek, C., and Garner, P. (2000), "Effects of treatment for intestinal helminth infection on growth and cognitive performance in children: Systematic review of randomized trials", *British Medical Journal*, 320: 1697–1701, and 321: 1224–330.

Dobbings, J., and Sands, J. (1978), "Head circumference, biparietal diameter and brain growth in fetal and postnatal life", *Early Human Development*, 2: 81–87.

Doucet, E., Almeras, N., White, M. D., *et al.* (1998), "Dietary fat composition and human adiposity", *European Journal of Clinical Nutrition*, 52: 2–6.

Dreze, J., and Sen, A. (1990), *Hunger and Public Action* (Oxford: Clarendon Press).

Dubowitz, L., and Dubowitz, V. (1977), *Gestational Age of the New-Born* (Menlo Park, Calif.: Addison-Wesley).

Easterlin, R., and Crimmins, E. (1985), *The Fertility Revolution* (Chicago: University of Chicago Press).

Ehrenberg, A. (1968), "The elements of lawlike relationship", *Journal of the Royal Statistical Society A*, 131: 280–302.

Elinder, L. (2005), "Obesity, hunger, and agriculture: The damaging role of subsidies", *British Medical Journal*, 331: 1333–6.

Falkner, F., Holzgreve, W., and Schloo, R. (1994), "Prenatal influences on postnatal growth: Overview and pointers for research", *European Journal of Clinical Nutrition*, 48: 15–24.

FAO/UNU/WHO (1985), *Energy and Protein Requirements*, Technical Report 724 (Geneva: World Health Organization).

Finkelstein, E., Fiebelkorn, I., and Wang, G. (2004), "State-level estimates of annual medical expenditures attributable to obesity", *Obesity Research*, 12: 18–23.

Fisher, R. A. (1935), *The Design of Experiments* (Edinburgh: Oliver and Boyd).

Flatt, J.-P. (1987), "Dietary fat, carbohydrate balance, and weight maintenance: Effects of exercise", *American Journal of Clinical Nutrition*, 45: 296–306.

Floud, R., Wachter, K., and Gregory, A. (1991), *Height, Health and History* (Cambridge: Cambridge University Press).

Fogel, R.W. (1994), "Economic growth, population theory and physiology: The bearing of long-term processes on the making of economic policy", *American Economic Review*, 84: 369–95.

Frazao, E. (1999), "High costs of poor eating patterns in the United States", in *America's Eating Habits: Changes and Consequences*, AIB 750 (Washington, D.C.: US Department of Agriculture, Economic Research Service), 5–32.

Freeman, J., Cole, T., Chinn, S., *et al.* (1995), "Cross-sectional stature and weight reference curves for the U.K., 1990", *Archives of Disease in Childhood*, 73: 17–24.

Gardner, G. W., Edgerton, V. R., Senewiratne, B., Bernard, J. R., and Ohira, Y. (1975), "Physical work capacity and metabolic stress in subjects with iron deficiency anemia", *American Journal of Clinical Nutrition*, 30: 910–17.

Georgescu-Roegen, N. (1966), *Analytical Economics: Issues and Problems* (Cambridge, Mass.: Harvard University Press).

Gibson, R. (1993), *Nutritional Assessment: A Laboratory Manual* (Oxford: Oxford University Press).

Glanz, K., and Rimer, B. K. (1995), *Theory at a Glance: A Guide for Health Promotion Practice*, NIH Publication 95–3869 (Washington, D.C.: National Institutes of Health).

—— Patterson, R., Kristal, A., *et al.* (1994), "Stages of change in adopting healthy diets: Fat, fiber, and correlates of nutrient intake", *Health Education Quarterly*, 21: 499–519.

—— Kristal, A., Tilley, B., and Hirst, K. (1998), "Psychosocial correlates of healthful diets among male auto workers", *Cancer Epidemiology, Biomarkers and Prevention*, 7: 119–26.

Goldberg, G., Black, A., Jebb, S., *et al.* (1991), "Critical evaluation of energy intake data using fundamental principles of energy physiology. 1. Derivations of cut-off limits to identify under-recording", *European Journal of Clinical Nutrition*, 45: 569–81.

Goldstein, H., and Thomas, S. (1996), "Using examination results as indicators of school and college performance", *Journal of the Royal Statistical Association*, A, 159: 149–63.

Gopalan, C., Rama, B., and Balasubramanian, S. (1970), *Nutritive Value of Indian Foods* (Hyderabad: National Institute of Nutrition).

Gorman, W. M. (1967), "Tastes, habits and choices", *International Economic Review*, 8: 218–22.

—— (1980), "A possible procedure for analyzing quality differentials in the egg market", *Review of Economic Studies*, 47: 843–56.

Grantham-McGregor, S. (1995), "A review of the studies of the effect of severe malnutrition on mental development", *Journal of Nutrition*, 125: 2233–8.

Gribble, J. N. (1993), "Birth intervals, gestational age, and low birth weight: Are the relationships confounded?", *Population Studies*, 47: 133–46.

Haas, J., Martinez, E., Murdoch, S., *et al.* (1995), "Nutritional supplement during the preschool years and physical work capacity in adolescent and young adult Guatemalans", *Journal of Nutrition*, 125: S1078–89.

Hamill, P., Drizd, T., Johnson, C., Reed, R., and Roche, A. (1977) "NCHS growth curves for children: Birth–18 years", *Vital and Health Statistics*, ser. 11, no. 165 (Washington, D.C.: United States Government Printing Office).

Harriss, B. (1995), "The intrafamily distribution of hunger in south Asia", in J. Drèze, A. Sen, and A. Hussain (eds), *The Political Economy of Hunger: Selected Essays* (Oxford: Clarendon Press).

Henderson, M. M., Kushi, L. H., Thompson, D. J., *et al.* (1990), "Feasibility of a randomized trial of a low-fat diet for the prevention of breast cancer: Dietary compliance in the Women's Health Trial Vanguard Study", *Preventive Medicine*, 19: 115–33.

Hobcraft, J., McDonald, J., and Rutstein, S. (1983), "Child spacing effects of infant and early child mortality", *Population Index*, 49: 585–618.

Houthakkar, H., and Taylor, L. (1970), *Consumer Demand in the United States*, 2nd edn (Cambridge, Mass.: Harvard University Press).

References

Hutchinson, R. (1969), *Food and Principles of Dietetics*, 12th edn (London: Arnold).

IFPRI (1990), *Research Report* (Washington, D.C.: International Food Policy Research Institute).

International Institute for Population Sciences (1995), *National Family Health Survey* (Bombay: International Institute for Population Sciences).

Ironmonger, D. (1972), *New Commodities* (Cambridge: Cambridge University Press).

James, W. P. T., and Schofield, E. (1990), *Human Energy Requirements* (Oxford: Oxford University Press).

—— Ferro-Luzzi, A., and Waterlow, J. C. (1988), "Definition of chronic energy deficiencies in adults", *European Journal of Clinical Nutrition*, 42: 969–81.

Johansson, L. (1972), *Production Functions* (Amsterdam: North-Holland).

Johnstone, A., Murison, S., Duncan, J., Rance, K., and Speakman, J. (2005), "Factors influencing variation in basal metabolic rate include fat-free mass, fat mass, age, and circulating thyroxine but not sex, circulating leptin, or triiodothyronine", *American Journal of Clinical Nutrition*, 82: 941–8.

Jones, P., Hunt, M., Brown, T., and Norgan, N. (1986), Waist–hip circumference ratio and its relation to age and overweight in British men", *Human Nutrition: Clinical Nutrition*, 40C: 239–47.

Joreskorg, K. (1970), "A general method for the analysis of covariance structures", *Biometrika*, 57: 239–51.

Kallberg, J., Jalil, F., Lam, B., and Yeung, C. (1994), "Linear growth retardation in relation to three phases of growth", *European Journal of Clinical Nutrition*, 48: 25–44.

Keusch, G., Fontain, O., Bhargava, A., *et al.* (2006), "Diarrheal diseases", in D. Jamison *et al.* (eds), *Disease Control Priorities in Developing Countries* (New York: Oxford University Press), 371–88.

Keys, A. (1980), *Seven Countries: A Multivariate Analysis of Death and Coronary Heart Disease* (Cambridge, Mass.: Harvard University Press).

—— (1984), "Serum cholesterol response to dietary cholesterol", *American Journal of Clinical Nutrition*, 40: 351–9.

Koenig, M., Foo, G., and Joshi, K. (2000), "Quality of care within the Indian family welfare programme: A review of recent evidence", *Studies in Family Planning*, 31: 1–18.

Kramer, M. (1987), "Intrauterine growth and gestational duration determinants", *Pediatrics*, 80: 502–11.

Kristal, A. R., Feng, Z., Coates, R. J., Oberman, A., and George, V. (1997), "Associations of race/ethnicity, education, and dietary intervention with the validity and reliability of a food frequency questionnaire", *American Journal of Epidemiology*, 146: 856–69.

—— Henderson, M. M., Patterson, R. E., and Neuhauser, M. L. (2001), "Predictors of self-initiated, healthful dietary change", *Journal of the American Dietetic Association*, 101: 762–6.

Kronmal, R. (1993), "Spurious correlation and the fallacy of the ratio standard revisited", *Journal of the Royal Statistical Society A*, 156: 379–92.

196

Kusin, J., Kardjati, S., Houtkooper, J., and Renqvist, U. (1992), "Energy supplementation during pregnancy and postnatal growth", *Lancet*, 340: 623–6.

Laird, N. M., and Ware, J. H. (1982), "Random effects models for longitudinal data", *Biometrics*, 38: 963–74.

Lancaster, K. (1971), *Consumer Demand: A New Approach* (New York: Columbia University Press).

Larsson, B., Svardsudd, K., Welin, K., *et al.* (1984), "Abdominal adipose tissue distribution, obesity, and risk of cardiovascular disease and death: 13 year follow up of participants in the study of men born in 1913", *Clinical Research*, 288: 1401–4.

Leibenstein, H. (1957), *Economic Backwardness and Economic Growth* (New York: John Wiley and Sons).

Leiberman, D. (1996), "How and why humans grow thin skulls: Experimental evidence for systemic cortical robusticity", *American Journal of Physical Anthropology*, 101: 217–36.

Levitsky, D., and Strupp, B. (1995), "Malnutrition and the brain: Changing concepts, changing concerns", *Journal of Nutrition*, 125: 2212–20.

LIMPDEP (1995), *LIMPDEP* (New York: Econometric Software Inc.).

Liu, K., Dyer, J., McKeever, J., and Mc Keever, P. (1976), "Statistical methods to assess and minimize the role of intra-individual variability in obscuring the relationship between dietary lipids and serum cholesterol", *Journal of Chronic Disease*, 31: 399–418.

Lozoff, B. (1988), "Behavioral changes in iron deficiency", *Advances in Pediatrics*, 35: 331–60.

Luby, S., Agboatwalla, M., Felkin, D., *et al.* (2005), "Effects of handwashing on child health", *Lancet*, 366: 225–33.

Manios, Y., Moschandreas, J., Hatzis, C., and Kafatos, A. (1999), "Evaluation of a health and nutrition education program in primary school children of Crete over a three-year period", *Preventive Medicine*, 28: 149–59.

Martorell, R., and Habicht, J.-P. (1986), "Growth in early childhood in developing countries", in F. Falkner and J. Tannen (eds), *Human Growth*, vol. 3 (New York: Plenum), 241–59.

—— and Scrimshaw, N. (eds) (1995), "The effects of improved nutrition in early childhood: The Institute of Nutrition of Central America and Panama Follow-up Study", *Journal of Nutrition*, 125. no. 4S.

Mincer, J. (1962), "Labor force participation of married women", in G. H. Lewis (ed.), *Aspects of Labor Economics* (Princeton, N.J.: Princeton, University Press), 63–106.

Mirrlees, J. (1975), "A pure theory of underdeveloped economies", in R. Reynolds (ed.), *Agriculture in Development Theory* (New Haven, Conn.: Yale University Press), 83–106.

Moe, C., Sobsey, M., Samsa, G., and Mesolo, V. (1991), "Bacterial indicators of risk of diarrhoel disease from drinking-water in the Philippines", *Bulletin of the World Health Organization*, 69: 305–317.

References

Monckeberg, F. (1975), "Effects of malnutrition on physical growth and brain development", in J. Prescott, M. Reeds, and D. Coursin (eds), *Brain Function and Malnutrition: Neurophysiological Method of Assessment* (New York: John Wiley), 15–52.

Monsen, E. R., and Balintfy, J. L. (1982), "Calculating dietary iron bioavailability: Refinement and computerization", *Journal of the American Dietetic Association*, 80: 307–11.

Monsen, E. R., Hallberg, L., Layrisse, M., *et al.* (1978), "Estimation of available dietary iron", *American Journal of Clinical Nutrition*, 31: 134–41.

Muhuri, P., and Preston, S. (1991), "Effects of family composition on mortality differentials by sex among children in Matlab, Bangladesh", *Population and Development Review*, 17: 415–34.

National Cancer Institute (2001), *5 a Day for Better Health Program*, National Institutes of Health Publication 01-5019 (Bethesda, Md.: National Institutes of Health).

—— (2006), "Pyramid Servings Database (PSDB) for NHANES III", available at <http://riskfactor.cancer.gov/pyramid/database/>.

National Center for Health Statistics (1977), *NCHS Growth Curves for Children: Birth–18 Years*, DHEW Publication PHS 78–1650. (Washington, D.C.: DHEW).

—— (2007), "National Health and Nutrition Examination Survey", available at <www.cdc.gov/nchs/nhanes.htm>.

Nelson, M., Black, A. E., Morris, J. A., and Cole, T. J. (1989), "Between-and-within subject variation in nutrient intake from infancy to old age: Estimating the number of days to rank dietary intakes with desired precision", *American Journal of Clinical Nutrition*, 50: 155–67.

Nestle, M., and Jacobson, M. F. (2000), "Halting the obesity epidemic: A public health policy approach", *Public Health Reports*, 115: 12–24.

Neumann, C., Bwibo, N., and Sigman, M. (1992), *Functional implications of malnutrition: Kenya project final report*, (Washington, D.C.: United States Agency for International Development).

Newey, W. (1987), "Efficient estimation of limited dependent variable models with endogenous explanatory variables", *Journal of Econometrics*, 36: 230–51.

Neyman, J., and Scott, E. (1948), "Consistent estimates based on partially consistent observations", *Econometrica*, 16: 1–32.

Numerical Algorithm Group (1991), *Numerical Algorithm Group Mark 13* (Oxford: Oxford University Numerical Algorithm Group).

Ounpuu, S., Woolcott, D., and Greene, G. (2000), "Defining stage of change for lower-fat eating", *Journal of the American Dietetic Association*, 100: 674–9.

Partnership for Child Development (2002), "Heavy schistosomiasis is associated with poor short-term memory and slower reaction times in Tanzanian school-children", *Tropical Medicine and International Health*, 7: 104–17.

Peters, D., Yazbeck, A., Sharma, R., *et al.* (2002), *Better Health Systems for India's Poor*, (Washington, D.C.: World Bank).

Philips, L. (1974), *Applied Consumption Analysis* (Amsterdam: North-Holland).

198

Piaget, J. (1987), *The Essential Piaget*, ed. Gruber, H. E. and J. J. Vonèche (New York: Basic Books).

Pinstrup-Andersen, P. (1988), *Food Subsidies in Developing Countries* (Baltimore: Johns Hopkins University Press).

—— (2002), "Food and agriculture policy for a globalizing world: Preparing for the future", *American Journal of Agricultural Economics*, 84: 1201–14.

Pitt, M. (1983), "Food preferences and nutrition in rural Bangladesh", *Review of Economics and Statistics*, 65: 105–14.

—— and Rosenzweig, M. (1985), "Health and nutrient consumption across and within farm households", *Review of Economics and Statistics*, 76: 212–23.

Pollak, R. (1969), "Conditional demand functions and consumption theory", *Quarterly Journal of Economics*, 83: 70–78.

Pollitt, E. (1993), "Iron deficiency and cognitive function", *Annual Review of Nutrition*, 13: 521–37.

—— Gorman, K., Engle, P., Martorell, R., and Rivera, J. (1993), *Early Supplementary Feeding and Cognition*, Monographs of Society for Research in Child Development, 58: 7.

—— and Schurch, B. (eds) (2000), "Developmental pathways of the malnourished child", *European Journal of Clinical Nutrition*, 54 (suppl. 2).

Popkin, B., and Doak, C. (1998), "The obesity epidemic is a worldwide phenomenon", *Nutrition Reviews*, 56: 106–14.

—— Armstrong, L., Bray, G., *et al.* (2006), "A new proposed guidance system for beverage consumption in the U.S.", *American Journal of Clinical Nutrition*, 83: 529–42.

Prais, S., and Houthakkar, H. (1955), *The Analysis of Family Budgets* (Cambridge: Cambridge University Press).

Prentice, A., and Jebb, S. (1995), "Obesity in Britain: Gluttony or sloth?", *British Medical Journal*, 311: 437–9.

Prentice, R., Chlebowski, R., Patterson, R., *et al.* (2006), "Low-fat dietary pattern and risk of invasive breast cancer: The Women's Health Initiative Randomized Controlled Dietary Modification Trial", *Journal of the American Medical Association*, 295: 629–42.

Preston, S. (1976), *Mortality Patterns in National Populations* (New York: Academic Press).

Pritchett, L. (1994), "Desired fertility and the impact of population policies", *Population and Development Review*, 20: 1–55.

Prochaska, J., and Di Clemente, C. (1983), "Stages and processes of self-change in smoking: Towards an integrative model of change", *Journal of Consulting and Clinical Psychology*, 5: 390–5.

—— —— (1984), *The Transtheoretic Approach* (Homewood, Ill.: Dow-Jones-Irwin).

Prochaska, J., Norcross, J., and Di Clemente, C. (1994), *Changing for Good* (New York: William Morrow and Co).

Pudney, S. (1981), "Instrumental variable estimation of a characteristics model of demand", *Review of Economic Studies*, 48: 417–33.

References

Purves, D. (1989), *Body and Brain: A Trophic Theory of Neural Connections* (Cambridge, Mass.: Harvard University Press).

Rand Corporation (1983), *Measurement of Physiologic Health for Children*, 5 vols (Santa Monica, Calif.: Rand Corporation).

Ravallion, M. (1990), "Income effects on undernutrition", *Economic Development and Cultural Change*, 38: 489–516.

Ravens, J. (1965), *The Coloured Progressive Matrix Test* (London: Lewis).

République Rwandaise (1986), *Enquête nationale sur le budget et la consommation des ménages* (Rwanda: Direction Générale de la Statistique).

Resnicow, K., Jackson, A., Wang, T., *et al.* (2001), "A motivational interviewing intervention to increase fruit and vegetable intake through black churches: Results of the eat for life trial", *American Journal of Public Health*, 91: 1686–93.

Rexrode, K. M., Carey, V. J., Hennekens, C. H., *et al.* (1999), "Abdominal adiposity and coronary heart disease in women", *Journal of the American Medical Association*, 280: 1843–8.

Reynolds, K., Franklin, F., Binkley, D., *et al.* (2000), "Increasing the fruit and vegetable consumption of fourth-graders: Results from the High 5 project", *Preventive Medicine*, 30: 309–19.

Rivera, J., Martorell, R., Ruel, M., Habicht, J.-P., and Haas, J. (1995), "Nutritional supplement during the preschool years influences body size and composition of Guatemalan adolescents", *Journal of Nutrition*, 125: S1068–77.

Rivers, D., and Vuong, Q. (1988), "Limited information estimators and exogeneity tests for simultaneous probit models", *Journal of Econometrics*, 39: 347–366.

Robinson, J., and Godbey, G. (1997), *Time for Life: The Surprising Ways Americans Use Their Time* (University Park, Pa.: Pennsylvania State University Press).

Rogoff, B., and Wertsch, J. V. (eds) (1984), *Children's Learning in the Zone of Proximal Development* (San Francisco: Jossey-Bass).

Rolls, B. J., Bell, E. A., Castellanos, V. H., *et al.* (1999), "Energy density but not fat content of foods affected energy intake in lean and obese women", *American Journal of Clinical Nutrition*, 69: 863–71.

Rosenstock, I. M., Strecher, V. J., and Becker, M. H. (1988), "Social learning theory and the health belief model", *Health Education Quarterly*, 15: 175–83.

Ryan, J., Bidinger, P., Rao, N., Pushpamma, P. (1984), *Determinants of individual diets and nutritional status in six villages of south India* (Hyderabad, India: International Crops Research Institute for Semi-arid Tropics).

Sahn, D. (1988), "The effect of price and income changes on food-energy intake in Sri Lanka", *Economic Development and Cultural Change*, 36 (2): 315–40.

Sahota, P., Rudolf, M., Dixey, R., *et al.* (2001), "Randomised controlled trial of primary school based intervention to reduce risk factors for obesity", *British Medical Journal*, 323: 1–5.

Schoefield, W. N. (1986), "Predicting basal metabolic rate: New standards and review of previous work", *Human Nutrition: Clinical Nutrition*, 39C (suppl. 1): 5–41.

Schroeder, D., Martorell, R., Rivera, J., Ruel, M., and Habicht, J.-P. (1995), "Age differences in the impact of nutritional supplement on growth", *Journal of Nutrition*, 125: S1051–9.

Schultz, T. W. (1961), "Investment in human capital", *American Economic Review*, S1: 1–17.

—— (1974), "Fertility and economic values", in T. W. Schultz (ed.) *Economics of the Family* (Chicago: Chicago University Press), 3–22.

Schutz, Y., Flatt, J-P., and Jequier, E. (1989), "Failure of dietary fat intake to promote fat oxidation: A factor favoring the development of obesity", *American Journal of Clinical Nutrition*, 50: 307–14.

Scrimshaw, N. S. (1996), "Nutrition and health from womb to tomb", *Nutrition Today*, 31: 55–67.

—— and SanGiovanni, J. (1997), "Synergism of nutrition, infection, and immunity: An overview", *American Journal of Clinical Nutrition*, 66: 464S–77S.

—— Taylor, C. E., and Gordon, J. E. (1959), "Interactions of nutrition and infection", *American Journal of Medical Sciences*, 237: 367–403.

Scrimshaw, S. (1978), "Infant mortality and behavior in the regulation of family size", *Population and Development Review*, 4(3): 383–403.

Sen, A. (1983), "Economics of the family", *Asian Development Review*, 1: 14–26.

—— and Sengupta, S. (1983), "Malnutrition of rural children and the sex bias", *Economic and Political Weekly*, 18: 855–64.

Shepard, T. Y., Weil, K. M., Sharp, T. A., *et al.* (2001), "Occasional physical inactivity and a high-fat diet may be important in the development and maintenance of obesity in human subject", *American Journal of Clinical Nutrition*, 73: 703–8.

Sheppard, L., Kristal, A., and Kushi, L. (1991), "Weight loss in women participating in a randomized trial of low-fat diets", *American Journal of Clinical Nutrition*, 54: 821–8.

SIFPSA (1996), *Performance Indicators for the Innovations in Family Planning Services Project*, State Seminar Report (Lucknow, India: Lucknow University Press).

Sigman, M., Neumann, C., Jansen, A., and Bwibo, N. (1989), "Cognitive abilities of Kenyan children in relation to nutrition, family characteristics and education", *Child Development*, 60: 1463–74.

Simon, H. (1986), "Rationality in psychology and economics", *Journal of Business*, 59: S209–24.

Smith, A. M., Baghurst, K., and Owen, N. (1995), "Socioeconomic status and personal characteristics as predictors of dietary change", *Journal of Nutrition Education*, 27: 173–81.

Smith, R., and Blundell, R. (1986), "An exogeneity test for a simultaneous equation Tobit model with an application to labor supply", *Econometrica*, 54: 679–85.

Smith, S. R., de Long, L., Zachwieja, J. J., *et al.* (2000), "Concurrent physical activity increases fat oxidation during shift to a high-fat diet", *American Journal of Clinical Nutrition*, 72: 131–8.

Sommer, A. (1986), *Nutritional Blindness* (New York: Academic Press).

References

Southon, S., Wright, A., Finglas, P., *et al.* (1994), "Dietary intakes and micronutrient status of adolescents: Effects of vitamins and trace element supplementation on indices of status and performance in tests of verbal and non-verbal intelligence", *British Journal of Nutrition*, 71: 897–918.

SPSS (1999), *SPSS for Windows Version 10* (Chicago: SPSS, Inc.).

Spurr, G. B. (1983), "Nutritional status and physical work capacity", *Yearbook of Physical Anthropology 1983*, 1–35.

STATA (2003), *Stata version 8* (College Station, Tex.: STATA).

Steptoe, A., Doherty, S., Kerry, S., Rink, E., and Hilton, S. (1999), "Sociodemographic and psychological predictors of change in dietary fat consumption in adults with high blood cholesterol following counseling in primary care", *Health Psychology*, 19: 411–19.

Stigler, G. (1945), "The cost of subsistence", *Journal of Farm Economics*, 27: 303–14.

Strauss, J., and Thomas, D. (1998), "Health, nutrition and economic development", *Journal of Economic Literature*, 36: 766–817.

Svedberg, P. (1990), "Under-nutrition in sub-Saharan Africa: Is there a gender bias?", *Journal of Development Studies*, 26: 469–85.

Tanner, J. (1986), "Growth as a target seeking function", in F. Falkner and J. Tanner (eds), *Human Growth* (New York: Plenum), 167–79.

—— Whitehouse, R., and Takaishi, M. (1966), "Standards from birth to maturity for height, weight, height velocity, and weight velocity", *Archives of Disease in Childhood*, 41: 454–71.

Taylor, C., Newman, J., and Kelly, N. (1976), "The child survival hypothesis", *Population Studies*, 30: 262–78.

Thomas, D., and Strauss, J. (1997), "Health and wages: Evidence on men and women in urban Brazil", *Journal of Econometrics*, 77: 159–85.

Timmer, C. P., Falcon, W., and Pearson, S. (1983), *Food Policy Analysis* (Baltimore: Johns Hopkins University Press).

Trussell, J. (1988), "Does family planning reduce infant mortality? An exchange", *Population and Development Review*, 14: 171–8.

UNICEF/UNU/WHO/MI (1999), *Preventing Iron Deficiency in Women and Children: Background and Consensus on Key Technical Issues and Resource for Advocacy, Planning and Implementing National Programmes* (Boston: International Nutrition Foundation).

UNICEF/WHO (1999), *Prevention and Control of Iron Deficiency Anaemia in Women and Children* (Geneva: World Health Organization).

United Nations (1992), *Child Mortality Since the 1960s: A Database for Developing Countries* (New York: United Nations).

Urban, N., White, E., Anderson, G. L., Curry, S., and Kristal, A. R. (1992), "Correlates of maintenance of a low-fat diet among women in the Women's Health Trial", *Preventive Medicine*, 21: 279–91.

USDA (1999), *Promoting Healthy Eating: An Investment in the Future*, Report to the US. Congress (Alexandria, Va.: Food and Nutrition Service.

—— (2003), *Continuing Survey of Food Intakes of Individuals* (Beltsville, Md.: Agricultural Research Service).

USDA/DHSS (1990), *Nutrition and Your Health: Dietary Guidelines for Americans* (Washington, D.C.: Department of Health and Human Services).

—— (2000), *Nutrition and Your Health: Dietary Guidelines for Americans*, 5th edn (Washington, D.C.: Department of Health and Human Services).

van Derslice, J., and Briscoe, J. (1995), "All coliforms are not created equal: A comparison of the effects of water source and in-house water contamination on infant diarrheal disease", *Water Research*, 29: 1983–95.

Vella, F., and Verbeek, M. (1999), "Two-step estimation of panel data models with censored endogenous variables", *Journal of Econometrics*, 90: 239–63.

Vygotsky, L. S. (1987), "Thinking and speech", in R. W. Rieber and A. S. Carton (eds), *Collected works of L.S. Vygotsky*, Vol 1. (New York: Plenum), 39–285.

Waterlow, J. C. (1986), "Metabolic adaptation to low intakes of energy and protein", *Annual Review of Nutrition*, 6: 495–526.

—— (1994), "Causes and mechanisms of linear growth retardation", *European Journal of Clinical Nutrition*, 48: 1–4.

—— Buzina, R., Keller, W., *et al.* (1977), "The presentation and use of height and weight data for comparing the nutritional status of groups of children under the age of 10 years", *Bulletin of the World Health Organization*, 55: 489–98.

Watkins, W. E., and Pollitt, E. (1997), " 'Stupidity or worms': Do intestinal worms impair mental performance?", *Psychological Bulletin*, 121: 171–91.

Willett, W. (1998a), "Is dietary fat a major determinant of body fat?", *American Journal of Clinical Nutrition*, 67: 556S–62S.

Willett, W. (1998b), *Nutritional Epidemiology*, 2nd edn (Oxford: Oxford University Press).

Williams, C., Strobino, B., Bollella, M., and Brotanek, J. (2004), "Cardiovascular risk reduction in preschool children: The 'Healthy Start' Project", *Journal of the American College of Nutrition*, 23: 117–23.

World Bank (1991), *Pakistan Integrated Household Survey* (Washington, D.C.: World Bank).

—— (1997), *Living Standards Survey Vietnam* (Washington, D.C.: World Bank).

—— (2005), *World Development Indicators* (Washington, D.C.: World Bank).

World Health Organization (2006), "Global Database on Body Mass Index", available at <www.who.int/bmi/index.jsp>.

Yen, S., Lin, B.-W., and Smallwood, D. (2003), "Quasi- and simulated-likelihood approaches to censored demand systems: Food consumption by food stamp recipients in the United States", *American Journal of Agricultural Economics*, 85: 458–78.

Young, L., and Swinburn, B. (2002), "Impact of the *Pick the Tick* food information programme on the salt content of food in New Zealand, *Health Promotion International*, 17: 13–19.

Subject Index

Note: page numbers in **bold** refer to Tables.

216

Author Index